Humanity Reimagined

Humanity Reimagined

WHERE WE GO FROM HERE

Martin Fiore

 **Rivertowns**
BOOKS

Printed in the United States of America · October 2021 · I

Cloth edition: ISBN-13: 978-1-953943-06-4
Paper edition: ISBN-13: 978-1-953943-05-7

LCCN Imprint Name: Rivertowns Books

Rivertowns Books are available from Amazon, B&N.com, and other online merchants as well as from bookstores and other retailer stores. Requests for information and other correspondence may be addressed to:

Rivertowns Books
240 Locust Lane
Irvington NY 10533
Email: info@rivertownsbooks.com

Contents

Putting People First

UNDERSTANDING TECHNOLOGY in all its forms has been a lifelong passion for me. For most of that time, it was an interest satisfied through reading. Then, as transformative technologies began to play a larger role in business, it became part of my professional specialization to advise clients on the benefits of new technology and its impact on business strategy, process, and operations. As computers and automation became the tools enabling increases in globalization and productivity, efficiency, and accuracy, I devoted much of my time and energy to understanding new developments in software and computing power, and I watched as information techno-logy transformed how business got done, driving economic expansion and growth worldwide. It seemed clear to me that accelerating technological innovation was a positive, constructive force in our society that would enhance the future of humanity.

By 2015, I had taken on a new role leading talent at Ernst & Young LLP (EY), which included internal learning, recuiting, and retaining new professionals, mobility, compensation and benefits programs, and other people-related strategies. Accordingly, I shifted my focus to the implications of advancing technologies for people both in and out of the workplace. I explored the new skills that our professionals would need to stay relevant and to realize the promise of new technological innovations, from information technology to artificial intelligence (AI). I believed that technological progress would give people labor-saving tools that would enable them to spend more time on critical thinking, strategic analysis, and following their purpose.

I also believed there would be a continuing need for uniquely human capabilities like imagination, empathy, and compassion that would allow people to simultaneously amplify their own contributions while also providing the human input needed to shape the guiding conscience for the new developments touching every aspect of life. Technology doesn't automatically support human values like ethics, equality, respect, and trust; the machines we build are value-neutral and can be used in ways that honor our values or violate them. Thus, the task of designing and embedding guardrails is left to us, the people who will envision and craft the technologies of the future.

I began speaking to business groups, university audiences, gatherings of colleagues, and young pro-fessionals about technology's potential impact on the future of work. Each year, the new trends needed updating and tweaking as the mind-boggling achievements mounted. New business models were emerging, from digital market platforms to ridesharing, homesharing, and crowdsourcing. New inventions were appearing, like electric and autonomous vehicles, drones, and 3D printing. And once unimaginable concepts like blockchain technology, cryptocurrencies, and quantum computing were moving from the margins of human imagination to take center stage.

What's more, technology began increasingly to converge with humans. No longer merely a support tool that resided in our homes or offices or even on our wrists, technology was *in* us—embedded, implanted, modifying our neural activities, our bodily processes, our genetic codes. We reached a different kind of tipping point for humanity. For the first time, the essence of both social and scientific transformation was inside us. And with that momentous turn, I felt we had reached a new level of extraordinary innovation—and disruption. This sense was only deepened by the arrival of the global COVID-19 pandemic in 2020, which demonstrated in a very tangible way that both the threat and the solution are in us.

The consequences of this shift will be momentous. Where earlier innovations impacted workforce policies, social interaction, and lifestyle options, many future changes will involve internal tweaking in the form of edited genetic code, installation of organ implants, and monitoring systems to guide our diets, fitness regimens, and mental activities.

We live in a world of remarkable possibilities. The pandemic of 2020-21 reminded us that natural forces can powerfully disrupt human societies. But the will to

preserve humanity through science and global collaboration stepped up in an awe-inspiring and positive
response. Hundreds of millions of people discovered that
they could use digital tools to remain productive and
interconnected even while thousands of miles apart.
Vaccines developed in record time using new mRNA
engineering techniques promised to dramatically slow or
even halt the ravages of COVID-19. And other new dreams
continued to emerge. Civilian space launches suggest a
future in which earthlings might choose to settle down in
another part of our solar system, or one in which major
forces behind earth-bound challenges like climate change
might be relocated to distant worlds. I believe the current
rate of technological change will lead to similarly seismic
shifts, and will do so in an even shorter time.

These are bold, exciting, and energizing possibilities.
Who wants to put a cap on human ingenuity? But the
continuing flood of technological changes raises the
overarching question: Are we asking all the right questions
to become prepared for breakthrough discoveries, unfore-
seen changes, and transformative events we don't even
have terms for yet? Who is refereeing serious developments that have social, economic, political, and philo-
sophical implications? Who can act as arbiter of the best
humanistic judgments? Are we watching out for both the
prosperous and the marginalized? Are we putting people
first? Disruption and innovation aren't forces beyond our
guidance; they are levers for advancement that humanity
itself should be controlling for positive benefits for all.

I am committed to raising awareness of these crucial
questions, and encouraging our readiness to act. It's a

commitment that has been deepened over the past six years by my responsibility as the father of two daughters. This responsiblity and blessing excites me about where we are going as a global society and about the future unfolding for all our children. It strengthens my resolve to help create a world where technology supports and improves us and the society at large—a place where we solve big problems like climate change, extend human life, and bring prosperity within the reach of all.

Our goal should be to manage the future disruptions that are surely coming with the preservation of humanity as our core objective. And because the scope of the issues we face is so broad and varied, no single person, country, industry, or organization can address them alone. We will need to unite our efforts in an ecosystem that brings together all the necessary components of assessment, solution, and process to create a system of shared beliefs regarding how we can maximize humanity as well as a set of common guidelines for doing so.

This book looks at where we are now, an era in which we will experience the greatest disruption and transformation of what it means to be human since the birth of civilization. In these pages, we'll explore the trends that hint at where we are headed.

Chapter one takes us quickly through the history of modern innovation, using the stories of past and present industrial revolutions as a gauge of important milestones. In chapters two and three, we look at the evolution of major forces—the unrelenting flow of time and progress— that have driven all these changes, and the trends that are pressing us forward in today's world. These trends are subject to continual change, of course, but exploring what is happening today can help us see around the next corner to glimpse what today's capabilities may signal for the future.

Finally, in chapters four through six, we look at what we can do as individuals, organizations, and as a species to bend the arc of history toward putting people—all people—first. Along the way, I shine a spotlight on individual case studies that illuminate some of the ways in which humanity is being reimagined and suggest the questions we must ask to make sure the changes under way will enhance rather than damage our human gifts.

I'm convinced that a fruitful journey toward the best future outcomes for humankind will start with forming an ecosystem of stakeholders from business, academia, government, nonprofits, communities, and, most important, individuals like you and me to raise awareness of our options, facilitate debate, and inspire wise, concerted action. My hope is that insights found in these pages will alert and enlighten you, inspire and challenge you. Together, let us embrace innovation with the goal of maximizing and enhancing our shared humanity.

Martin Fiore
September 2021
New York City

CHAPTER ONE

Human and Technology Convergence in the Era of the Fifth Industrial Revolution

WHEN WE SEEK TO EXPLORE the potential impact of technology on humanity and society, it's helpful to look to history to help us understand our past evolution and provide insights into where we might be headed. The influences that have shaped human reality include socio-economic trends, lifestyle changes, population migrations, public policies, and the introduction and adoption of inventions such as the steam engine, electricity, manufacturing assembly lines, the telephone, indoor plumbing, the automobile, the airplane, the computer, and, most recently, intelligent automation and autonomous systems.

Focusing particularly on the impact of technological innovations, many experts have characterized the broad sweep of recent human history as including a series of major shifts, often referred to as industrial revolutions (see Table 1-1).

Table 1-1: The Five Industrial Revolutions			
Revolution	Starting Date	Duration	Core Change
First Industrial Revolution	1765	105 years	Transition from hand production to machine tools
Second Industrial Revolution	1870	99 years	Introduction of mass production and process standardization
Third Industrial Revolution	1969	31 years	Birth of computerization and the information age
Fourth Industrial Revolution	2000	18 years	Digitization, intelligent automation, and artificial intelligence
Fifth Industrial Revolution	2021	Unknown	Convergence of humankind and machines

Even a brief glance at Table 1-1 gives a vivid impression of how the tools and technologies mentioned have deeply and distinctively impacted human life. It's also easy to notice the decreasing time span between important eras and the increasing sophistication of the developments, with each period building on the ones before.

The first four industrial revolutions shown in the table are widely recognized by historians. Today, this continuing process of technological evolution and its accelerating pace are moving us toward a pivotal point at which humans are increasingly converging with machines. This change, driven by the literal fusion of humans and technology, is the one that we have listed as the Fifth Industrial Revolution. It involves a growing symbiotic relationship that joins the human body and mind with technologies as well as with the environment in which we live and from which we derive our continued existence.

This new relationship between humans and technology is reflected in such familiar advancements as digitally connected fitness devices, including bracelets and watches that track workout activities and biomedical data. There also are more recent developments in the area of connected holistic wellness, such as implanted and consumed medical tracking devices—for example, digestible sensors and brain-computer interfaces that monitor and interact with neural networks. New developments in these areas enable the gathering of biomedical data, such as blood sugar, blood pressure, pH balance, body mass index, and hormone levels. The data can readily be shared with multiple practitioners involved in an individual's care, including one's physician, nutritionist, trainer, or therapist. Eventually, these new forms of tracking technology may combine with science and education to create holistic wellness models customized to the needs of each user. Beyond merely tracking performance, these models could guide human behaviors, predicting dietary, medicinal, or exercise needs.

New tools and systems are also being developed that enable human physical capabilities to be augmented and enhanced for use in activities from manufacturing and space exploration to military and police work. Technological ingenuity is improving human capabilities in a host of ways, from new devices that assist and support physical rehabilitation to increasingly sophisticated and powerful forms of prosthetic limb design.

One remarkable recent example is the advent of *industrial exoskeletons*, suits worn on the factory floor that use pneumatics or hydraulics to amplify the worker's strength and performance while reducing the risk of work-related injuries. These techno-garments keep workers safer while boosting productivity and improving quality. Other forms of wearable robotics are being developed to help people suffering from paralysis and limited mobility, including, for example, the more than 70 million people worldwide who experience brain injuries and strokes each year. Still other emerging technologies are addressing human-body complexities in ways that were never thought possible, such as replicating intuitive motor control, touch sensation, and extremity manipulation for the additional millions who are affected by amputations, severe burns, and other forms of debilitating physical trauma.

These are all examples of how humans and technology are becoming increasingly linked in physical ways. At the same time, the convergence of humans with technology is happening on an intellectual and psychological level. This convergence is reflected in the increasing use of artificial intelligence and machine learning to augment the human capacity for gathering, analyzing, interpreting, and using information to solve problems and make better decisions.

It's hard to overstate the importance of this trend. As far back as we can trace, the human brain has been the most powerful tool for making sense of the world. Now that age-old reality is being upended by a new paradigm— a world in which the most powerful sense-making tool is the human brain operating as part of a dynamic network (partly biological, partly electronic) that can tackle challenges of incredible complexity with far greater speed, thoroughness, and accuracy than ever before possible.

Addressing the changes driven by the Fifth Industrial Revolution is perhaps the biggest challenge faced by the organizations of today and tomorrow, from for-profit businesses and nonprofit organizations to government agencies, educational institutions, and more. It's also the core subject of this book.

Robotic Process Automation— A Glimpse of the Convergence to Come

THE RISE OF ROBOTIC process automation (RPA) is a good starting point for the human-technology convergence story and its continuing evolution in contemporary organizations. It demonstrates how technology can support and replace human effort, a realization that has sparked reactions that run the gamut from excitement and hope to fear. Paradoxically, the story of RPA also illustrates the fact that most businesses have still made only limited investments in time-tested productivity drivers.

Despite what the name might imply, an RPA tool is more akin to a software program than to a futuristic machine with metal limbs. In many cases, an RPA "soft bot" is simply a tool that mimics the keystrokes of a human operator. Such soft bots come in handy whenever you have a process of information gathering and distribution—pushing data and pulling data. Automation of the manual steps involved in such processes can eliminate many hours of human effort and increase both efficiency and quality. Thus, RPA has the power to replace or supplement human labor by relieving people of arduous, repetitive tasks and freeing them up for more interesting and fruitful challenges that involve higher-end skillsets and human-to-human roles. I call this shift "taking the robot out of the human."

In the early 2000s, RPA was a hot topic among companies seeking to optimize efficiency and improve the quality of many back-office functions, such as extracting data from documents, filling out forms, and generating mass mailings.

These activities are not dramatic or glamorous. But what RPA lacks in star power, it makes up for in practical, nuts-and-bolts delivery value. It does a lot of work that most humans really do not want to do, and it does it faster and better. That may sound like a backhanded compliment, yet RPA has proven to be an important gateway tool to higher forms of artificial intelligence (AI) as well as to the relief of worker tedium.

Thus, introducing an intelligent automation technology solution like RPA into the work environment is about shining a light on fragmented processes and activities carried out by humans that needed to be fixed.

To achieve the best long-term result, it's important to integrate the humans and the bots, so that they learn to work together rather than against each other. A simple example might be an RPA application that reduces delays in a business's automated processes by recognizing a bottleneck, identifying an employee with the power and position to remedy it, and then figuratively "tapping them on the shoulder" to remind them that the next step in a process is ready to be performed.

RPA is a simple example of the kinds of technologies, designed to replicate human thought and actions, that are fundamentally altering how we work. Some fear this trend as a threat to human jobs. But technology doesn't necessarily take away jobs; it cuts back on *tasks,* freeing up and supplementing human colleagues so the humans can focus on what they do uniquely well: relationship building, analysis, innovating, and creating.

Thus, employees who understand the benefits of using RPA bots appreciate the value they add to the process. They offer a way to free up employees to learn new skills that benefit them and the company, such as Six Sigma methods, coding, or Design Thinking—yielding both a much leaner, less expensive process and a better equipped (and potentially happier) employee.

As the example of RPA suggests, advances in AI and machine learning are enabling organizations to improve quality, performance efficiency, and overall productivity. One study projects that half of today's work activities across all industries could be automated by 2055 or earlier. While the study concludes that fewer than five percent of all occupations are likely to be automated entirely, for many occupations about one-third of all activities may be automated.

Furthermore, occupational patterns will change dramatically under the impact of automation over the next few decades. For example, there are likely to be more computer technicians and far fewer retail store salespeople, given both the increase in technology dominance in developed societies and the continuing growth of online shopping. The long-term shift in labor force activities over the next couple of decades has been compared to the move away from agriculture and toward manufacturing during the First and Second Industrial Revolutions.

More recent generations tend to "get this" right away and are more open to embracing new technologies. Bots and machine learning programs don't threaten their jobs, they enhance them. Innovative momentum, powered by the gradually declining cost and increasing availability of technology, coincides with greater demand on humans to be more agile and to add more value. In this sense, new technologies can be effective workplace and community partners, opening a whole new world of possibilities for what people do with their time. For employees, this could mean machines deal with the *what* while people deal with the *why*—assessing, analyzing, advising.

However, the degree to which industries have been automated using tools like intelligent automation and AI is still limited. Costs, labor market dynamics, and social acceptance factors all play a role in the relatively slow rate of adoption for these technologies. However, process automation did increase during the COVID-19 pandemic of 2020-21 as companies took steps to automate workflows in response to temporary worker shortages. Thus, the RPA industry, which was valued at $250 million in 2016, had an estimated 2021 value of just under $3 billion.

Intelligent automation tools have proven to be just the first step in what has continued to be an evolving journey. This journey is driving business executives, consumers, social scientists, educators, and policy makers to think more strategically about how to capitalize on technology tools while achieving some level of equilibrium with our collective commitment to humanity—in the form of workers, employees, students, friends, and family members.

The Accelerating Pace of Change and the Rise of AI Autonomy

ONE OF THE WAYS WE CAN SEE that the coming convergence of humans and machines is not just another technological trend but actually the harbinger of a new industrial revolution is the increasing pace of change. The series of industrial revolutions humankind experienced beginning in the 18th century involved the development of increasingly powerful machines that have played a steadily greater role in our everyday lives. By the end of the 20th century, almost everyone in what are called the developed nations of the world had a lifestyle in which ordinary human capabilities were dramatically enhanced by the capabilities of machines. Our ability to move about the planet, once dependent on the physical strength of our bodies, was enhanced by devices like automobiles and airplanes; the reach of our five senses was enhanced by tools like televisions and computers.

Now the process of merging with digital technologies and inviting them into our lives has become even more intimate. In some cases, we've literally embedded nanotechnology inside our bodies to monitor our heart rates or to help fight diseases like cancer. We've adopted AI in the form of virtual assistants that keep millions of households informed and entertained. Robotic vacuums have become routine household cleaning appliances. Sales of plug-in electric vehicles topped three million in 2020. In one field after another, new technologies are making our work faster or easier, strengthening our connections with other people, and generally enhancing our lives and our future prospects.

Novelist Ernest Hemingway once wrote that a man goes broke "gradually and then suddenly." Something similar might be said about the ways our lives are being changed by technology. The quickening pace of change is matched only by the scope of impact that comes with it. It's no longer just about incremental improvements, it's about once unimaginable possibilities—in our tools, within our bodies, and for our society. Some of the changes we're likely to experience will feel almost miraculous. The blind will see, intractable diseases will be cured, genetic codes will be reshaped. But these changes mean that humanity also is facing a pivotal moment fraught with potential challenges.

Developments in AI and supercomputing are moving inexorably toward the literal convergence of people and technologies, making it possible for intelligent machines to occupy our bodies and act on our behalf. One result is that it is increasingly possible for these machines to act as they see fit with little or no human consultation.

For decades, such a shift didn't seem feasible. Computers were remarkable pieces of hardware that used binary code to do complex calculations in the blink of an eye—but few people believed that computers could ever displace more uniquely *human* characteristics such as creativity, empathy, and insight. They couldn't write the great American novel, make complex business decisions, or exhibit compassion for a person in need. However, with the extraordinary power of quantum computing on the horizon, the potential for breaching those barriers is no longer the far-fetched stuff of science fiction.

Today we're seeing more examples of intelligent machines acting autonomously, having been trained by their inventors to both imitate and reflect human capabilities. They can mirror our movements, our decision patterns, and even our habits. They can drive us around, order our favorite takeout, measure our heart rates, and automatically regulate our insulin levels after checking our blood sugar. They can also beat chess masters at their own game, write passable poetry, and conduct dialogues with depressed or troubled people that sound remarkably like those a trained therapist would engage in.

These are amazing abilities—but they give many people pause. When the devices that impact us are given power to act autonomously, it's understandable that some people may find this disquieting—more so than when machines act simply in partnership with humans and under their direct control.

In the 2004 film *I Robot*, the protagonist asks a robot whether it can compose a beautiful symphony or paint a masterful canvas. The robot turns the question back on the human: "Can you?" At the time, this exchange effectively made the distinction between machine capabilities and human creative superiority. But today, it's reasonable to assume that with access to the right technologies, both characters could reply with an emphatic "Yes!"

Machine intelligence will play an increasingly important role in our society as humans and technology move toward convergence. It will continue to transform how and where we work, where we live, and how we engage with each other. It will alter how we make decisions, which skills we learn, and how we run our economies. Yet as the influence of machine learning grows, there still is very little public discussion about where it's all headed—or how we should get there.

Perhaps we have become so comfortable with the impact of technology that we don't even register each new development. Some would argue that we've effectively ceded control and put machine learning in the driver's seat. The most visible outcome may be the potential impact on jobs. But there's also a much deeper, more profound reengineering process happening around how technology is affecting what it means to be human, from our value systems to our connections with each other.

There may be some prophecy or larger lesson to be taken from a description of the potential benefits and risks of the industrial exoskeleton authored by a team of experts from the Centers for Disease Control and Prevention. "There is potential for over-reliance on the technology," the analysts warned. They explained that it's possible that workers using an exoskeleton could end up assuming that dangers like muscle strain, nerve damage, and other risks are no longer to be taken seriously, perhaps resulting in misuse and long-term damage to humans. "Before the widespread implementation of industrial exoskeletons occurs," the experts said, "research is needed to evaluate their effectiveness in reducing the risk factors." In this field as in most other areas of innovation, it's prudent to remember the law of unintended consequences.

Will we commit to prudently evaluating the risks of convergence, including those caused by over-reliance on new technologies? Will we carefully examine the space between technological disruption and the enormous opportunities it brings for humanity and society, studying the ramifications and making reasoned choices before simply ceding control to the devices we've created? Will we be intentional in taking actions to preserve and maximize the best of what it means to be human?

The fact is that technology is disrupting the status quo in irrevocable ways. To survive and thrive in this period of unrelenting and unprecedented change, it is to our advantage to keep up and not give up. The more we understand about what is happening and work to acquire the skills and mindset needed to help define how it affects us, the better we'll be able to help shape and advance our society in a humanistic way.

The Social Context for Today's Technological Disruptions

THE TECHNOLOGICAL CHANGES I've briefly sketched are not happening in a vacuum. Instead, they are unfolding against a backdrop of social and economic changes that makes the potentially disruptive impact of the new technologies even more powerful. Several specific changes are worth noting.

First, today's younger generations—true digital natives, born after 1980 and raised in a technology environment—are moving into roles across all industries that put them in a position to invent, develop, market, and buy new technologies and applications. Their comfort level with machines inspires bolder use of technological tools in life and work. They're also exploring ways to use these powerful tools to address social and environmental issues ranging from climate change to economic inequality. As a result, skepticism about new technologies is steadily shrinking.

Second, humanity has never been more poised for positive change. Our era of rapid transformation may be filled with uncertainty, but it is also one of the most exciting times in human history. Despite many serious economic and social challenges, humanity has never been smarter or healthier. The fields of science never have been more aligned around addressing profound issues such as life improvement and extension, ecosystem and climate balance, resource husbandry, and exploration beyond our known world. Thanks to digital technology advances, education and information are more accessible than ever.

And, on average, businesses and, yes, even some governments are more fiscally fit. These resources give humankind the potential to apply the new technologies in ways that are especially creative and beneficial.

Third, we've entered a new period in which humans around the globe are more deeply interconnected than ever before. In 2020-21, the COVID-19 pandemic dramatically demonstrated this new reality. (See the sidebar on pages 26-28 for my summary of the major social impacts of the pandemic.) The spread of the virus from human to human so rapidly across countries and continents demonstrated human connectivity tangibly and on a massive scale. Even as we experienced the pandemic in deeply personal ways, we also became more aware of the vulnerabilities we share and the importance of working together to ensure healthier times by listening to science and making the right choices.

As COVID-19 reminded us, with all the advancements we've made and the new technologies available, there still is a raw human element that overpowers and takes precedent above all else in times of shared crisis.

Five High-Impact Outcomes
of the COVID-19 Pandemic

During 2021, as the world began to emerge from the worst months of the COVID-19 pandemic, it became clear that some aspects of life would be changed for a long time to come. World wars and financial crises have yielded similarly momentous changes, of course, but the COVID pandemic felt uniquely disruptive, partly because it was both global and intensely local and partly because of the sheer numbers of lives lost and the inability of those left behind to grieve according to long-held traditions.

Here's my take on five lingering impacts of the pandemic.

- *We got comfortable with being uncomfortable.* In April 2021, it was reported that, for the first time, the most popular emoji on Twitter was a face with mouth open and tears streaming, a reflection of sheer emotional overload. Many felt pushed to their limits more than once during the pandemic, which represented an assault so pervasive and complex that many—perhaps for the first times in their lives—felt a powerful and sustained lack of control over even the simplest and most mundane life activities. The upside: The pandemic provided a unique opportunity for millions to learn to live with and manage discomfort, which many psychologists call the secret sauce for a higher achieving, more satisfying life.
- *Talent and capital became more dispersed.* In mid-July 2021, the *Wall Street Journal* ran a story with the headline "Innovation Moves to Middle America," highlighting

a pandemic-driven shift of business resources away from traditional U.S. technology centers and into smaller, more affordable cities like Tulsa, Oklahoma City, and Phoenix. One possible consequence: The move of entre-preneurs and tech innovators to the exurbs may give them exposure to a wider range of "real-world chal-lenges" and therefore potentially more humanity-driven innovations in activities like healthcare and farming, transportation and housing.

- *We rediscovered the urgent need for education reform.* The school shutdowns and the shift to virtual learning exposed long-standing problems in the U.S. educational system, especially the unequal treatment of students from varied economic and social backgrounds. The good news is that the crisis has given educators and adminis-trators a rare opportunity to upend traditional thinking and start fresh, thinking not only about how to promote learning effectively through times of isolation but also about how to make school systems more responsive to the needs of all children. And this is a global issue. Rec-ommendations from experts at UNESCO, the UN's edu-cational agency, include establishment of distance learning programs; emergency planning around school closures and home-based quarantines; providing the tools and resources that students at all socioeconomic levels require; and ensuring that students with disabili-ties or from low-income backgrounds have equal access to digital devices, internet connectivity, and other sup-port systems.

- *Many of us have embraced the hybrid model of work.* Months of working remotely have had a lasting effect on the work styles of millions of people, especially highly skilled employees and business professionals. The pandemic broke through cultural and technology obstacles that had previously slowed the wide implementation of remote work options. One study estimates that just over 20 percent of the U.S. workforce could work remotely three to five days a week as effectively and productively as they would in an office. The number could be even higher in economies with a different occupational and industry mix—for example, the UK, where business and financial services are a larger share of the economy. The long-term impacts of a shift toward hybrid work may be huge and varied. In 2020, New York City reported 15,000 vacant rental apartments, the most in recorded history. At the same time, remote workers may shift their spending patterns: while clothing sales may drop, home office equipment purchases may rise.

- *We realized afresh how interconnected we are.* If there was any lingering doubt about the smallness and fragility of our collective life on earth, the pandemic erased it. Under pressure from COVID, geographic boundaries, global economics, and geopolitical forces were all shown to be relatively weak and vulnerable, highlighting the mutual dependencies of our healthcare systems, natural resources, and supply chains. Lessons learned? Probably too many to count, but one key insight is the need to deliberately develop an international ecosystem to prepare for future global health emergencies. After all, the next pandemic is very likely incubating somewhere right now.

We witnessed a similar shift recently in the waning years of the Fourth Industrial Revolution. The people/machine equation moved away from people using technology and toward people and technology symbiotically using one another to achieve mutually desirable outcomes. By 2021, we were looking at how we could leverage technology to help us get through to the next (and hopefully final) chapter of the global COVID pandemic—tracking and predicting outbreaks, keeping businesses going with effective remote working, contact tracing, monitoring and analyzing emerging virus variants and, most crucially, deploying and delivering vaccines. Could technology be the key enabler to save humanity and society from future global pandemics through better projections, modeling, planning, and rapid response?

Fourth, there has been a gradual loosening of institutional influence on long-held norms around codes of conduct as well as around everyday ethics, privacy, and trust. From educational and legal systems to religions and governments, traditional guideposts and guardrails no longer hold the sway they once did in influencing personal and institutional allegiance to these norms. The rise of social media, the decline in religious affiliation, widespread information-sharing, and general lowering of privacy thresholds have all played a role in making humans more vulnerable to possible misuse of technology and data gathering.

Finally, countless people are avidly seeking technological fixes for the biggest problems we face. We are 13 billion years into the life of our known universe—only the last

200,000 of which have seen the evolution of modern humans. And despite many medical breakthroughs in the past two centuries, it has taken more than 1,000 years just to double human life expectancy. With all the amazing technology at our disposal today, isn't it natural we would be getting a little impatient for awe-inspiring innovations?

The answer is yes, but the question remains: At what cost? When we speak of maximizing humanity and the betterment of human existence, what does that really mean? Despite the accelerating pace of technological change, it's still up to us to dictate the ground rules.

Aligning People and Technology:
Four Imperatives

SO HOW DO WE PUT PEOPLE FIRST when it comes to the convergence of humans and technology? How will we humans get along with and collaborate with our technology-based partners to create the best possible combination of economic and social progress? What are the basics for adapting and succeeding in this new landscape?

In the face of this challenge, I consider myself a "transformation optimist." I believe that humankind can and will put people and their needs before technology, which I believe is essential to realizing outcomes that are more *Star Trek* than *Terminator*, as author Jerry Kaplan has described it.

Applying this approach requires that we develop an ecosystem of leaders from every sector—business, academia, government, nonprofit organizations, faith groups, and communities—who are committed to working together to ask and answer the big questions about where we are headed and how we want to get there. This ecosystem can ensure we stay connected, communicate often, hear the voices of the marginalized as well as the powerful, and move along parallel tracks rather than pursuing strategies that will lead to conflict and confusion. The people and organizations that do this best are the most likely to transform the populations they manage and govern. People must continue to lead the transformation, acting as the fulcrum for the technology lever. A positive approach can feed the pipeline of transformative ideas. Instead of being victims of unprecedented change, we can seize on this time to be both innovators and empowered thinkers.

As we navigate this time of disruption and uncertainty, leaders must be transparent with their constituencies and stakeholders, and work with them to foster skillsets that will continue to be relevant in times of change. As younger people from the Millennial generation and Gen Z become a larger part of the population, we can appeal to their generational characteristics of being naturally collaborative, comfortable with framing a point of view, and decisive. There are a few clear competencies that stand out as must-haves for more recent generations—analytical reasoning, problem solving, innovative mindset, and project management capabilities. These are capabilities that not only set them apart in job searches or leadership roles but also clearly differentiate them from their digital coworkers.

I believe there are four critical principles we all need to practice as we seek to engage others in addressing the challenge of technological transformation and successfully aligning it with our basic human values:

- *Strive for agility.* Change is hard. But getting ahead of change, finding clarity in the churn of enormous amounts of information, and achieving consensus around the way forward is necessary to steer through the fog and inspire constructive coping. I believe technological innovation is moving us away from hierarchical structures and toward more democratic means of managing our world and society. Digitally savvy younger generations are driving these changes as they bring new approaches and attitudes to the workplace and to our communities. Flexibility and mobility are important cultural traits for them, and the same qualities lend themselves to faster adaptation of new ideas.

- *Invest in knowledge.* In order to define and defend the uniquely human values that should shape our future, we must have a deep understanding of our place in the universe and the ways in which the shifting tides of change have brought us to our current situation. This kind of knowledge can't simply be downloaded; it's gained through continuous learning, experience, mentorship, and a broad sense of history. It is crucial that it be combined with tech tools and algorithms to help us realize what it will take to guide meaningful change. Machines look at data. They spew out numbers, patterns, and calculations, but can't yet understand all the reasoning behind the numbers. We humans are differentiated and empowered by our ability to combine quantitative data with qualitative lessons and observations to produce actionable insights and strategy.

- *Leverage our human connections.* Robots—at least for now—don't have feelings. They can't connect with other machines or people in a meaningful way. We need to foster and use our innate and distinctive human ability to forge meaningful relationships. Team up with people you regard as cutting-edge thinkers and build out your network to get broad exposure to new ideas. As we expect more from technology, we must not expect less from each other. Cultural analyst Sherry Turkle, who has studied how our tech devices and online personas are redefining human connection and communication, argues that nothing can substitute for true human understanding. With the uptick in digital communications, leaders have a responsibility to support human connection, making a conscious effort to balance "high tech" and "high touch."

- *Encourage digital proficiency.* Productivity skyrockets when we integrate domain knowledge and people skills with machine-proficient processes. As we gain new skills and acclimate to evolving technologies, productivity will increase, actually lifting employment and contributing to income parity. We must prepare now for a future in which artificial intelligence will play an increasingly dominant role, rather than being surprised by the scope and impact of those changes. The future requires a seamless arc between our early awareness, our increased knowledge and understanding, and our ability to innovatively leverage technology for the best possible solutions for humankind.

Our goal should be to embrace this time of unprecedented possibilities and use it as a catalyst for positive change. That means putting people first. I believe it is possible for us to thrive in a time of disruption and transformation while also ensuring we ignite action aimed at the betterment of humanity and society. Let's start the conversation.

The Challenge We Face
Is Not a Zero-Sum Game

SHAPING OUR TECHNOLOGICAL FUTURE should not be viewed as a zero-sum game, in which the capacities and powers gained by machines must come at the cost of human beings. Nor should it become a competitive race for dominance that will benefit a relative handful of highly trained individuals while impoverishing millions of those who are less well prepared. It should be about improving life for all in a manner that is thoughtful and just. It should be about seeing hopeful possibilities and then creating the opportunities those possibilities represent.

It requires us to come to a shared understanding of what is better, to passionately seek that result, and then to radiate positive outcomes in ways that benefit the most people in the most ways.

Imagine if we framed our approach to every challenge by saying, "I see a possibility . . . " Let's create a vision of a better reality and set the stage for understanding and action directed at the desired result. Let's practice seeing the world as it is while projecting the destination we seek.

We can all agree that enabling sight in the blind and eradicating poverty are desirable goals. If there are technological solutions that help us accomplish those goals, we should all agree on the value of pursuing them. And there are hundreds more examples of ways to improve human existence that would also gain near-universal agreement. For example, most of us would likely agree that unconscious and discriminatory bias in computer programming or algorithm development is not something we want to permit, nor do we want to allow technology investments to be channeled toward terrorism or nuclear war. There are many other goals we could all agree on quickly if we focus on making choices that will lead to an optimized human existence.

But not all such choices are clear-cut. We may get a majority to agree that autonomous vehicles can play an important role in transportation and package delivery options. But the question becomes much less black and white when malfunctioning circuitry causes a fatal accident, or when the guidance system in an autonomous vehicle must decide in a split second whether to swerve to avoid a head-on collision, thereby saving the car's passenger—but taking the life of an innocent nearby pedestrian.

How do we manage the awesome new possibilities that technology is creating so that uniquely human characteristics are preserved and stay in charge? It starts with embedding crucial human values into the design, development, and implementation of technological advances. Doing this will enable us to shape the narrative, putting humans at the center. Over time, we can become naturally predisposed to putting the best interests of humanity and society ahead of efficiency, productivity, convenience, and financial gain.

The risks of not managing this properly are huge. We are facing the most monumental set of decision points humanity has ever experienced. In the midst of rapid technological advancement, we are also wrestling with a global health pandemic and economic and social justice crises. The collective impact is forcing us to reflect on what needs to change and how we want to move forward as a society.

If we choose to sit back and let these events simply unfold, failing to take an active role in how we move forward, we may ultimately pay a steep price. We will continue to advance in the short term, but eventually we will create deeper divides, contributing to greater inequalities and leaving future generations in a worse place.

On the other hand, we can take action to preserve and maximize the best of what it means to be human by taking well-reasoned risks, investing in innovation, and grappling with the touchy questions regarding how we can bend the arc of history in humanity's favor. We also need to decide which values and principles we most want to protect and use those as waypoints to guide technology's inexorable march forward. I believe the values and principles of trust, ethics, sustainability, privacy, and learning can serve as those waypoints. They can be the framework we use in our discussions and debates, and in our efforts to engage more people in this cause.

As we face the possibilities presented by ever-more-sophisticated artificial intelligence and the yet-unknown capabilities of quantum computing, it is time to ask a few critical questions:

- Are we moving too fast?
- How will we build and instill trust in technology and each other in a society with more access to information and fewer boundaries?
- How do we fruitfully connect the powers of people and technology, and then harness the combination for the benefit of all humankind?

It all comes back to one basic premise: In this era of unrelenting technological transformation, we must raise awareness, facilitate debate, and be intentional in our actions to preserve and maximize the best of what it means to be human. In the remaining chapters of this book, we'll explore the implications of this premise in more detail.

SPOTLIGHT

Windows on the World—How Brain-Computer Interfaces Are Restoring and Enhancing Human Capabilities

Scientists estimate there are over 40 million people worldwide who are completely blind. Most have problems involving injuries to the retinas in the eyes, but some suffer from damage to the nervous pathways that run between the eyes and the brain.

Restoring sight to individuals in the latter category is a matter of repairing the eye-brain connection rather than fixing something wrong with the eye itself. This is the focus of the work of Eduardo Fernández, director of the Neuroengineering and Neuroprosthesis Unit at the Bioengineering Institute at Miguel Hernández University in Spain.

Based on the principal that the human nervous system functions much the same as an electrical device, Fernández conceived a solution to the problem that would enable light and form to take shape in a person's mind by sending electrical signals directly to the brain. He spent years creating a brain-computer interface (BCI) system that records the environment of the blind person, then uses electrodes implanted in the brain to deliver very-low-resolution images straight to the patient's visual cortex. The process can be compared to plugging a camera into the video-cable port of a TV monitor—in this case directly into the patient's brain.

The idea of using electronic tools to augment the capabilities of the human body isn't new. The first cardiac pacemaker, designed to generate a regular electrical pulse that stimulates the natural rhythm of the heart, was implanted in a patient way back in 1958. The concept of cochlear implants that help to restore hearing to the deaf dates back almost as far, though modern versions of these devices didn't arrive on the scene until 1977. Fernández's work on applying a similar kind of technology to overcome blindness is the latest development in this arena.

Brain-computer interfaces (BCIs) like the one Fernández has developed are truly the stuff of science fiction brought to life. The work of neuroengineers using BCIs to cure, support, or enhance life experiences is a powerful example of the positive side of human-machine convergence, one with vast potential for the future. And the underlying science behind BCIs is fascinating in its own right.

Fernández first had to figure out what kind of electronic signal the human retina produces. He used organs from human cadavers to test what happens when retinas are exposed to light and how electronic pulses are sent and received. Since human retinas can be kept alive only for about seven hours, he then turned to machine learning to match the retina's electrical output after receiving a visual message.

For the next step, Fernández inserted a kind of camera prosthesis into the brain to deliver the signal or the visual message. One of the first patients using Fernández's system was a 57-year-old woman who had lost her sight at age 42. Surgically equipped with Fernández's combination camera, implant, and video feed device, she was able to reopen a window on her world, seeing very low-resolution dots and shapes providing the basic outlines of a physical object like a door or sidewalk. Facial recognition was more complex, but over time Fernández augmented his system with image recognition software that now helps the patient identify people up close. Through experimentation, he uses each new AI innovation to strengthen his solution.

Eventually, Fernández and others will create a prosthesis, similar to a cochlear implant, that can transmit its signals through the skull itself rather than an implanted device inside the skull. It is unknown how long such an implanted device can be left in the brain without causing damage, so the doctor and his team will have to figure out just how much stimulation the human brain can take without over-stressing the nervous system and ruining the quality of the images being received. But MRIs conducted on the first patient after having the device implanted for up to six months did not show any damage.

As with all highly complex breakthroughs of this kind, this effort will require great care and constant reevaluation. There's no question, however, that this work has been a huge step forward when it comes to human and machine convergence as well as new hope for the blind.

In the years to come, a range of new BCIs will be using artificial intelligence to train machines to learn some very complex and uniquely human capabilities. These include reading emotions, feeling empathy, and achieving direct communication between humans and computers using brainwaves and eye movements. BCI is helping solve many sensory and cognitive problems by enabling humans to use the signals produced by their brain activity to interact with and change their environments. For example, BCIs are already playing a critical role in helping people disabled by neuromuscular disorders such as stroke, spinal cord injury, and cerebral palsy to replace or restore some lost functionality of movement, speech, or writing using tools like robotic limbs, mechanized wheelchairs, and computer keyboards.

In July 2021, the *New England Journal of Medicine* marked a related step in reporting on new BCI technology that may one day help people "speak" by simply thinking what they want to say. University of California researchers working with a man in his thirties who had lost his speech to a stroke used this technology to record his brain activity as he observed individual words displayed on a screen and imagined saying them out loud. The words captured were accurate just under 50 percent of the time—but the accuracy level rose to 76 percent when researchers incorporated word-prediction algorithms similar to those used in email and texting. This new technology may soon offer hope to the thousands who lose the ability to talk each year due to illness or injury.

It seems clear that the evolving science of BCIs will help to generate a range of new insights into the workings of the human nervous system as well as many practical benefits. These are wonderful blessings for humankind. But it's up to all of us—leaders and practitioners in the scientific and medical communities, government regulators, academic experts, and society as whole—to be thoughtful about the development and use of BCIs to make sure that the benefits are shared in the most equitable fashion. We need to think about questions like:

- How can we encourage scientists and researchers to focus their BCI-related research on challenges with the greatest potential benefit for humankind as a whole rather than focusing primarily on problems that promise the biggest short-term economic benefit—for example, BCIs that might enhance the capabilities of high-salaried professional athletes?

- How can the intellectual property developed through BCI research, from basic biological discoveries to the designs of breakthrough devices, be made available widely to encourage widespread adoption and continuing scientific progress?

- How can we ensure that the people most desperately in need of medical remediation via BCIs will get access to the support they need regardless of their economic circumstances or social status?

Answers to questions like these may help to determine whether we look back on the advent of BCIs as one of the miracle breakthroughs of human history or—less hopefully—as the birth of a new category of "luxury goods" from which only a small segment of humankind is able to benefit.

CHAPTER TWO

Beyond Technology: The Three Key Forces Driving Change

A S WE SAW IN CHAPTER ONE, technology is changing our lives in countless ways, from how we collaborate with colleagues at work to how we identify and cure diseases, find the perfect vacation spot or the most compatible mate, and so much more. But technology is not the only factor driving major disruption, nor is it the only factor giving rise to questions about the future of humanity.

The most important driving forces we need to consider when mapping the future of humankind are long-term, powerful, dynamic movements that penetrate every aspect of civilized life, relentlessly reshaping the nature of our existence and largely fueled by human behavior. People can't control the passage of time, but human beings, individually and collectively, are continually making choices about how their time is used. People didn't create the

Earth, but we have embraced activities that lead us to cross its geographic barriers; to explore and exploit its natural resources; and to develop and spread cultures, patterns of behavior, and commercial and governing authorities that impact our global environment in powerful ways. People didn't invent the principles of natural variation, competition for resources, and survival of the fittest that define the workings of biological evolution, but the ways we choose to respond to those realties shape the demographic and generational shifts that are molding the future of humankind.

In this chapter, I want to focus on three key driving forces that are shaping today's reality in ways that that profoundly affect our day-to-day lives: globalization; the accelerating pace of innovation; and demographic and generational shifts.

Globalization: One World, Increasingly Interconnected

ADVANCES IN TRANSPORTATION SYSTEMS and telecommunications infrastructures have contributed immeasurably to the spread of knowledge, exchange of ideas, trade, cultural mingling, and understanding. In 2000, 95 percent of the largest global companies—those with revenue of $1 billion-plus—were headquartered in the so-called developed nations of North America, Europe, and East Asia. But some experts project that before the end of the 2020s, nearly half of the world's largest companies will be headquartered in developing nations, scattered throughout Asia, Latin America, and the Middle East. Economic activity will spread with them.

One result of increasing globalization is greater urbanization and exposure to different lifestyles and ways of thinking, contributing to generally enhanced tolerance, though certainly not eliminating pockets of hostility between groups defined by ethnicity, religion, and national origin. Globalization is opening economic and diplomatic doors and helping to fuel growth and change in emerging markets. It is giving us more choice as consumers, making us broader thinkers, and contributing to the proliferation and spread of innovative ideas, from new technologies to novel approaches to healthcare.

In 2000, the International Monetary Fund attempted to define globalization by identifying its four main characteristics: (1) trade and transactions; (2) capital and investment movements; (3) migration and movement of people; and (4) dissemination of knowledge. Many public and private organizations promote these pillars of international cooperation and growth. But the fund's analysts also noted a schism in how globalization was viewed around the world. For some, it was a highly beneficial process—"a key to future world economic development," as the IMF wrote at the time. For others, globalization threatened to increase economic inequality and jeopardize progress made on social and environmental issues.

Twenty years later, we've seen firsthand the benefits of globalization. Advances in transportation systems and telecommunications infrastructure have contributed to the spread of knowledge and exchange of ideas around the world. They have boosted trade, increased cultural mingling, and improved cultural understanding. A greater exposure to different lifestyles and ways of thinking has generally enhanced tolerance for different cultures and perspectives.

Unfortunately, the legacy of globalization also includes serious issues, such as climate change, potential threats to the safety and sustainability of our food production systems, increasing pressures on supplies of pure and potable water, the ease with which viral diseases and other health threats can spread across national borders, the rise of massive human migrations, and the threat of long-term conflicts exacerbated by all these physical and social challenges. Yet despite these problems, the global community has never been more unified. It is important that the spirit of economic and cultural integration be brought to bear on maximizing humanity for the long term.

Regardless of whether globalization is perceived as good or bad, it was clear in 2000—and remains clear now—that the process of globalization won't be stopped. Our job is to ensure it is managed for the good of all people, not just a fortunate few.

The Accelerating Pace of Innovation: The Future Is Coming, Ready or Not

IN CHAPTER ONE, I BRIEFLY ALLUDED to the accelerating pace and scope of technological change. Now I want to take a deeper dive into this phenomenon and its implications for humankind today.

The ability to imagine something new and make it real and useful has been a characteristic of human society for thousands of years. Innovation is usually sporadic and fairly slow. Yet periodically, through the long march of human existence, major, rapid innovative breakthroughs have occurred with enormous social and economic impacts.

Consider, for example, Johannes Gutenberg's invention of the printing press in the 15th century. The invention had so much impact on social systems that the term "Gutenberg Moment" has come to define a phenomenal development capable of exerting paradigm-shifting influence on human life for long periods of time. Think atomic bomb, space travel, the digital computer.

Like those later singularly powerful innovations, over time, the printing press exerted an even greater overall influence than the original purpose of the invention itself might suggest. Gutenberg spent years cobbling together a series of inventions—movable type, adjustable machine molds, oil-based inks—which eventually facilitated mass production of printed material.

What truly differentiated Gutenberg's invention from earlier book-printing processes was the speed with which his press could produce many copies of a manuscript. It is said that Gutenberg's first print run of the Bible in Latin, for example, took three years to produce 200 copies. By modern standards, of course, this is extremely slow. But in comparison with traditional methods, it represented a remarkable productivity breakthrough. The ability to produce multiple copies of books comparatively quickly meant that information could be shared much more rapidly and with a much wider group of audiences.

It is doubtful that Gutenberg himself ever imagined the massive changes to society and institutions that would emerge from his humble press. But the ability to produce large quantities of printed information that could in turn be shared among hundreds, thousands, and later millions of people radically changed how messages were documented and shared. Gutenberg's machine democratized information and enabled knowledge sharing on an unprecedented scale. It led to a vast explosion of literacy in the developed countries of Europe and North America, and then throughout the rest of the world. Eventually, it enlightened entire societies about ideas once restricted to an educated elite, facilitating social movements and political revolutions that would reshape the future of humanity.

Gutenberg's invention has impacted our world for more than five centuries now. We've now come to see it as just one of a series of inventions that have transformed society.

The laying of the first transatlantic cable in 1858 enabled instantaneous communication between America and Europe, stirring cultural curiosity that in turn drove greater interest in travel as an accessible form of entertainment, education, and business opportunity. The next half-century saw a series of developments that fed this interest, including the construction of railroad networks across much of the world and the invention of the automobile and the airplane.

A wide range of technologies began to reshape American culture in the mid-twentieth-century, including enhancement of telecommunications components and the growing ubiquity of television. As with the Gutenberg press, TV provided a sharing experience that opened a new lens on how other people lived, worked, and thought. It held up a mirror to who we are and motivated the dreams and ambitions of Americans at every demographic level. The communication and transportation breakthroughs of the 19th and 20th centuries point to the role of enhanced speed in making innovation even more potent. Whether yielding efficiency, expanding reach to greater numbers of people, or just saving time, speed was becoming a key differentiating factor between a common development and an extraordinary one.

Perhaps the watershed moment of the entire 20th century was the development in 1947 of ENIAC, the first general-purpose, digital computer. It required a 50 x 30-foot space and its own air conditioning system, and it took days to reprogram between numerical problems to be solved. Despite these limitations, ENIAC was the most sophisticated "thinking machine" built up to that point. Originally developed for military purposes, it wasn't finished until World War Two was over, after which it played a role in the calculations related to construction of the hydrogen bomb.

The decades since then have seen a focus on making technology tools smaller and less expensive, vastly increasing their global reach. The massive size and heat generation of early computers rendered them impractical tools for most businesses. The invention of the transistor made it possible to shrink computers while also making them more powerful. In fact, the transistor made just about everything electronic smaller, including millions of every-day items like radios and TV sets.

Today's microprocessors still use transistors, but they are much smaller and far less expensive. Engineers developed ways to put more and more transistors on a single computer chip, the wafer-like integrated circuits that give computers their processing and memory capacity. In 1965, Gordon Moore, a research and development director at a major semiconductor company, predicted that the number of components, or transistors, on an integrated circuit would double every year. His extrapolation was based on the cost per component being inversely proportional to the total number of components—in other words, the more you added, the cheaper they got. By 1975, what came to be called Moore's Law had proven to be amazingly accurate. For perspective, the most advanced chips today hold nearly 50 billion transistors. A transistor-based version of ENIAC today could fit on a grain of sand.

The chip turned out to be an awesome, general-purpose technology—a modern equivalent of the steam engine or electricity—which meant it was a development that could affect an entire economy, lifting all boats and paving the way for many diverse innovations in a wide range of industries. In fact, leading economists have credited improvements in the efficiency and pricing of integrated circuits with one-third of all U.S. productivity growth since the mid-1970s. From smart phones to breakthroughs in genetic medicine, chips have empowered human ingenuity, enabling tiny tools to process massive amounts of data for an amazing range of activities.

Some now believe we have seen the end of Moore's Law—that we can no longer expect the same rate of progress in fitting more computer power into smaller and smaller spaces. Others argue that there will be new ways to keep doubling the power of the chip, such as 3D architectures and new transistor designs. But we may have seen the end of increasingly faster and cheaper chips at a rate of every couple of years.

At a minimum, most experts believe chip design will become much more specialized, focused on specific purposes such as deep learning AI applications. One startup born at the Massachusetts Institute of Technology is betting on a fundamentally different kind of chip specializing in the types of calculations needed to run powerful AI programs. It performs these calculations using light beams shining through microscopic channels, in which information is efficiently encoded on different light wavelengths.

While this new "light chip" is still experimental, big tech is watching closely, because it promises to deliver greater AI power with lower use of energy. Notably, its calculations are analog rather than digital, which makes it unsuitable for precision calculations but extremely useful in training large, deep neural networks that can use data to make decisions in areas like natural language understanding.

Whatever the future holds for the light chip, sometime in the next decade, someone will develop the next iteration of what we now view as the all-purpose chip. The new design is likely to have an impact on numerous technologies that we can only begin to imagine.

As this brief recap of the technological developments of the last century suggests, the pace of innovation has accelerated dramatically since the days of Gutenberg. One big reason is the spread of tools needed to develop and implement new technologies.

It used to be that the resources needed to take an idea from concept to reality were available to relatively few people. Thus, it was rare for new developments to proliferate far and wide.

In the years to come, the process of technological transformation is likely to happen at an even faster pace, helped along by such trends as the development of a global market for highly efficient and inexpensively manufactured chips, resulting in greater affordability and more widespread use.

But the availability of inexpensive processing power and connectivity is only one chapter of the story. Today many other resources needed to innovate are easily accessed and mostly low-cost, making innovative capabilities and outcomes more regular and more affordable. All it takes to turn some ideas into reality is a laptop, a 3D printer, and some basic design skills. The tools that used to serve as barriers to innovation are largely accessible and mostly low cost. This has made it easier and cheaper to create and think outside the proverbial box. As a result, innovation has become society's main disruptor, trend enabler, and shaper of products, services, systems, and processes.

Automating manufacturing plants once involved equipment, physical adaptations that took time to raise capital investments and required long implementation periods. By contrast, RPA software, for example, can be rapidly adopted. Minimal coding skill is required to automate the low-level digital work tasks formally performed by employees, whether business executives or front-line workers. This has made RPA an ideal gateway technology for enterprises beginning to think more broadly and more strategically about how to capitalize on emerging innovations. Today, higher levels of AI technology are required for an expanded range of activities, such as document reading, speech, and image recognition. And the momentum toward automation that has developed in recent years has produced a pace of change that is truly unique.

These developments in turn have helped to spawn technology-enabled business models, including a host of platform-based businesses. They've also made it possible for new ideas, products, and companies to attain scale with stunning speed, far more affordably than in the past.

As a result, we've almost become accustomed to barrier-busting, jaw-dropping innovations. We're numb to the sheer ubiquity of technology in our lives and the constantly increasing speed of change—lulled into expecting a Gutenberg Moment nearly every year or even more frequently than that. Satisfying any itch, solving any perceived problem, just can't happen fast enough. (Turn to pages 56-60 to see my forecast of ten Gutenberg moments that may be on the horizon for the next two decades.)

This is the phenomenon of innovation—how the pieces, the additive adjustments, and the one-thing-leads-to-another building blocks all come together, adding up, over time to produce a once-unimaginable difference, for good or for bad. The power of major innovations creates the urgent need to establish guardrails around the aspects of what it means to be human that we wish to protect.

This is why the accelerated pace of innovation is a second powerful trend shaping our future.

Innovation itself is also changing its nature. Technology has always supported and enabled human endeavors. But in recent years, technological innovation has shifted to not just enabling but augmenting human capability in all its forms. We saw some examples in chapter one, where we discussed the gradual convergence of humans and machines that a range of technologies is making possible.

That shift is obviously exciting—but it also offers a warning sign. We face the possibility, for example, that our current approach to artificial intelligence will pose increasingly greater challenges as AI becomes more pervasive and human-capable. We want smart technologies to help us fulfill our true and full potential as humans, but machines don't share many of the characteristics of what it means to be human. A robot doesn't have empathy, moral and ethical codes, or emotions. So while we want to leverage new technologies to help solve problems and make life better, we also need to confront the question of how to make sure machines that learn stay aligned with our needs.

As a result, we need to reckon with the changing relationship between time and innovation. Time once was "the great thickener" of untested ideas. Given enough time, a new idea would show its true value. That equation has shifted: As the space between consequential developments gets shorter, the amount of time allowed for careful assessment also has shrunk. That means the faster cadence of new developments leaves us with less time to carefully evaluate whether these innovations actually add value—or even pose an existential threat—to our lives.

I'm reminded of a conference I participated in at the University of Notre Dame in 2018, titled "Artificial Intelligence and Business Ethics: Friends or Foes?" A dozen speakers and panelists from professional firms, technology companies, and academia gathered in South Bend to explore the ethical implications of AI and share our thoughts on the best path forward. A key question we asked was: Is our only option to trust these new technologies to independently act in our best interest? It's a question that continues to be relevant today, and one we need to proactively address, since these technologies will continue to evolve regardless of whether we want them to.

It is safe to assume that most of us want smart technologies to help us fulfill our true and full potential as humans, but machines don't share many of the characteristics of what it means to be human, such as empathy, moral and ethical codes, and emotions. While we want to leverage new technologies to help us solve problems and make life better, how do we teach machines to improve life? How do we use technology to make things better, not just faster, easier, or cheaper?

The Next Gutenberg Moments

The acclerating pace of technological change we see today is only going to get faster. The rising rates of rare, so-called "black swan events" in business and economics, along with the accelerating pace of paradigm shifts in technology, unpredictable medical crises, and urgent social challenges, cause me to believe that the next two decades will feature an unusually large number of remarkable breakthroughs.

The COVID-19 pandemic of 2020-2021 may prove to be another driver of accelerated change. History shows that other great public health challenges, such as flu and yellow fever pandemics, served to inspire pivotal changes. The European cholera epidemic of the mid 19th century, for example, led to the development of a sewer system in London that greatly improved its urban environment and helped it grow into a thriving metropolis with global influence. In Paris, the same outbreak inspired urban planners to create wider boulevards and more parks, providing people with more open space and ultimately gifting the world one of its most beautiful and widely admired cityscapes. The COVID pandemic may prove to have a similar seminal impact.

Here are my predictions for 10 future developments that may qualify as humankind's next set of epochal Gutenberg Moments:

1. *All companies become technology companies.* Business leaders are rethinking everything, looking for opportunities to completely data-enable companies to transform their operating models, revise business processes, redesign value chains, and constantly upskill

their employees. The internet-based cloud, where virtually limitless amounts of data and processing power can be housed, is serving as the catalyst and the enabler of these changes. As a result, the timeline that defines how frequently technology skills need updating—historically around 50 percent every couple of years—is likely to be drastically accelerated. In addition, data ethics and privacy will become increasingly important issues for all companies, no matter what industry they compete in.

2. *All (or almost all) products become services.* Ownership of countless physical products—from cars and houses to clothing, appliances, furniture, and tools—will become optional in the near future as virtually everything we need for day-to-day life will be available to be rented, borrowed, or shared. More and more companies will generate income by selling subscriptions or memberships rather than by selling products.

3. *The global rate of violent crime will be drastically reduced.* Despite an uptick in homicide rates during the COVID pandemic, crime rates in the U.S. have plunged to historically low levels since peaking in the 1970s. There's reason to believe the rest of the world is likely to follow suit in the years to come—an urgent issue, since, according to the UN Office for Drugs and Crime, many more people worldwide die every day from interpersonal crime and organized violence than in war zones. Society will mobilize around reducing such preventable deaths through investments in improved protection, risk reduction strategies, and limiting the availability of weapons. According to the World Economic Forum, reducing lethal violence in just seven of the most violent

countries in Latin America could save more than 365,000 lives over the next 10 years.

4. *Mental illness and its consequences, such as suicide, will be greatly reduced.* This is another social problem that may be poised for positive breakthroughs on several fronts. Wearable devices will enable people to get mental as well as physical feedback on their levels of wellness and thereby help reduce suicide risks. Mental health apps will be capable of recognizing a change in a user's sociability level (through behaviors like sending fewer texts or moving around less) and responding with advice or counseling. Online tools paired with smartphone sensors, all driven by AI, can provide further help and timely interventions to those in need.

5. *Electronic sports will surpass traditional sports in popularity and economic impact.* In an era when professional sports stars earn many millions of dollars, video game champions are quietly beginning to rival them. In 2019, total winnings in the Fortnite Tournament were $30 million, with the top winner taking home a $3 million purse. Plans are in the works to expand the e-sports market to compete with traditional sports, with dedicated areans beginning to open up around the world.

6. *3D-printed buildings will become commonplace.* 3D-printed homes, stores, offices, and other structures are likely to become a standard product of the construction industry by the mid-2020s. Fast, flexible, and low-waste, 3D printing drastically reduces the amount of concrete used in construction, which will help cut the major share of greenhouse gas emissions produced by the global building industry.

7. *Traffic congestion will become largely obsolete.* While highway signs and street lights are here to stay, internet-connected vehicles and devices like smart traffic lights will soon start making it easier for drivers to get around while also improving the speed and convenience of public transportation. AI traffic-management tools that use sensors to collect data on moving vehicles have shown they can reduce waiting times at traffic lights by as much as 42 percent, which also helps reduce auto emissions. Such systems already are being tested in cities like Atlanta, Pittsburgh, and Portland, Maine. Other AI tools provide real-time traffic data to vehicles and drivers, enabling quick rerouting to avoid the worst congestion.

8. *Climate-change adaptation will be a core mission for builders, city planners, and government agencies.* With more than 90 percent of urban areas in coastal regions, it's projected that by 2050 more than 800 million city residents could be threatened by rising sea levels. Another 1.6 billion people could be vulnerable to extreme heat conditions (eight times the number in 2021), while 650 million could face water scarcity. Experts are working on a wide range of solutions to build resilience to these threats. Some are complex and costly, such as retrofitting infrastructure. Others are nature-based, like planting trees next to streets. Still others are policy-centric, like encouraging city managers to incorporate risk awareness and prepardness into existing processes, including financial and insurance planning and optimizing emergency responsiveness.

9. *Premature deaths from accidents and illness will plummet.* Self-driving cars will eliminate a significant

fraction of road accidents; preventive, monitored, personalized, and virtual medicine will improve the treatment of acute illnesses and make it available to populations that are currently underserved; and robotic instruments, computer implants, and 3D-printed organs will gradually improve the effectiveness of many surgical instruments and processes. Taken together, trends like these will sharply reduce the number of people who die before the age of 70, playing a big role in the increase in the average lifespan that most experts foresee in the years to come.

10. *Exploration of outer space will resume, beginning with travel to the Moon and Mars.* After a long hiatus, journeys beyond Earth will return to the agenda of humankind, this time driven largely by private organizations rather than national governments. The purposes of space travel will include scientific study, business and industrial research, and new forms of adventure tourism. Some are hoping that colonization of other worlds and perhaps the discovery of evidence of alien life will also become part of the story.

Demographic and Generational Shifts: The Changing Faces of Humankind

GLOBALIZATION IS CREATING once-impossible connections across nations, regions, continents, and cultures. The accelerating pace of innovation is providing us with previously unimaginable capabilities that can be used for good purposes or for dangerous ones. And even as these two forces are challenging humanity with a range of new questions, demographic and generational changes are also transforming the makeup of humankind itself. These changes are giving rise to continental migrations, transforming global markets, and impacting basic human services, such as healthcare and availability of clean air and water, while also shining a bright light on human rights.

Shifting patterns of population growth are one of the demographic trends that is transforming life on Earth. The relationship between population growth and economic growth has been widely documented and studied for at least 200 years. Today, we see low population growth in high-income countries creating social and economic issues, while significant population growth in many lower-income countries is slowing their economic development. The relationship between population and economic growth also plays into contemporary social and political issues around international migration and natural resource consumption. Economic growth is obviously crucial in raising standards of living, including access to healthy food, housing, medical care, and education.

A second important demographic trend is urbanization. More than half of the world's population now lives in urban areas, concentrated in what the World Economic Forum (WEF) calls "megacities" with populations of 10 million or more. Urban life offers many advantages, but it creates problems, too, as the COVID-19 pandemic reminded us. Beyond disease, urban areas face other major threats, including pollution, loss of green space, and loss of land mass due to climate change. As just one example of the challenges we face, the World Resources Institute estimates that the number of people living in water-scarce areas will grow from the current 1 billion to 3.5 billion by 2025.

Technology and human ingenuity can do a lot to address these problems. To reimagine humanity in the most positive way, it is increasingly critical that our current economic model starts to assess progress using social, environmental, and governance metrics as well as traditional economic and financial indicators. There are signs of hope. The United Nations Sustainable Development Goals and the Paris Agreement on climate change have provided a framework for action, mapping important shifts in the way global societies produce, consume, and operate, and encouraging close cooperation between the public and private sectors.

There are signs that this shift in attitudes is working. New Clark City in the Philippines, located about 100 kilometers from Manila, is a resource-efficient city that is being built from scratch. About two-thirds of the city will be reserved for parks, farmland, and other green spaces. All buildings will use the latest technology to reduce water and energy consumption. Similar projects are underway in India and South Korea.

From the private-sector perspective, a 2017 report by the Business and Sustainable Development Commission cites goals outlined by the United Nations in 2015 that point the way to making the world more sustainable by 2030. The programs recommended represent a $12 trillion opportunity for businesses in the key areas of food and agriculture, cities, energy, materials, health, and well-being. Providing further underpinning for this way of thinking, the Business Roundtable, an organization representing the CEOs of the largest companies in the U.S., expanded its defined purpose for a company in 2019 to include the notion of serving all stakeholders, which includes employees, customers, and local communities, as well as shareholders.

A rise in migration, spurred in part by an increase in the number of political and economic refugees fleeing from one country to another, is another demographic trend that is challenging humankind. Political changes inspired by the Arab uprisings in 2011 are still impacting ethnic and religious strife in the Middle East and North Africa region. The refugee crisis also has had a significant impact on regional stability, leading to violence, sectarian persecution, and economic hardship. The UN estimates that in 2019 there were 272 million international migrants globally, or about 3.4 percent of the world population, up more than half a percent since 2000. About one-seventh of the world's population has migrated at least once in their lives.

This growing trend is causing challenges for many of the countries involved. Africa's population, for example, is one of the youngest and fastest growing in the world. On the other end of the scale, Europe has the largest population percentage over 60. These types of disparities will cause economic and demographic problems. The motivation to migrate is driven by many factors, including the desire to pursue economic opportunities, the need to escape from crime and personal harm, and the quest for greater gender parity. When left unmanaged, migration movements can lead to shortages of skilled workers in some parts of the world while sparking struggles over food, housing, and education in others. Yet migration could offer strategic benefits to countries like Japan, Germany, and the United Kingdom if we can develop the ability to match migrant skills with national market needs.

Yet another major demographic change currently under way is the aging of the human population. Extended life expectancy is one of the most significant overarching outcomes of technological and scientific developments over the past century. In 1913, for example, life expectancy worldwide was 34 years. In 2000, the last time global statistics were registered, it was 67. And in 2018, for the first time ever, the number of humans over age 64 exceeded those under age five. Access to information, incalculable advances in medicines, public health practices, surgery, organ replacement, healthcare monitoring, human mobility—all of these trends are helping people experience longer, healthier, and more active lives.

But an aging population also brings problems. As the COVID-19 pandemic demonstrated, elderly populations tend to be more vulnerable to disease, particularly to new virus strains, overwhelming healthcare systems. And then there's the public cost of supporting a vastly larger, older population, stressing pension systems and national economies that need to be focused on a vast array of social challenges and priorities. The future of education and work are also impacted as the increase in life expectancy means that more old people are continuing to learn and to continue to work long past traditional retirement ages, creating an unprecedented generational age spread in many workplaces. Addressing the future of aging will be a challenge for leaders across all sectors of society—from healthcare innovation to infrastructure planning, financial services, autonomous vehicles, digital assistants, educational systems, and more.

Finally, from a generational standpoint, we also see big shifts in interests, attitudes, and preferences. These are seismic social changes, way beyond the usual differences in musical tastes, clothing styles, and lingo. Younger generations are digital natives—they were born into a world enlivened by technological transformation. For many Millennials and members of Gen Z, their bank is an app, and shopping is an online transaction that comes with home delivery. They are buying cars from vending machines and attending classes virtually, 24/7.

They are also gifted with expansive minds and mores. Gender identification is a personal matter and a choice, as is sexual orientation. Virtually any societal barrier that may have been viewed as sacrosanct in previous centuries has been transformed into an open invitation for disruption.

What's Next for Humanity? How the Three Driving Forces Interact

SOME MIGHT ARGUE THAT the three forces I've high-lighted are not the only ones that deserve to be examined, and perhaps not even the most important ones. For example, some might point to the rise of pandemics or climate change as driving forces that are more likely to determine what human life will be like 50 years from now. But I counter that it is greater human movement and global exposure that generates pandemics, while the rapid pace of medical innovation has the potential to prevent them. At the same time, global cooperation and a new generation's sensible approach to sustainability holds the promise of solutions to today's serious environmental threats.

What's more, these same three forces I've high-lighted—globalization, the pace of innovation, and demographic and generational shifts—can be linked back to earlier periods of transformation and advancement throughout modern history. To see how this works, let's return for a moment to the first big example of innovation that I cited in this chapter—the Gutenberg press. A major technological turning point in information-sharing, the Gutenberg press also represented the impact of globalization, connecting people and ideas across national borders through easily preserved and transported information tools.

As historian Ada Palmer describes it, the real value of Gutenberg's invention was realized when there was a distribution network for dissemination of the printed word. Literacy rates were quite low in the late 15th century, so in any single town there was a limit to the influence of a particular book or news pamphlet. But after Gutenberg, sailors heading for near or distant shores could take copies with them and sell them to local printers in their ports of call, who in turn could reprint and distribute more copies. Locals would gather in pubs or other town centers and listen as a paid reader relayed the content of the newest printed documents. In this way, information could spread across the world far more quickly than in the past, changing forever the nature of human connectedness.

From these beginnings, the first global news and information networks were launched. The new availability of books produced quickly and inexpensively in multiple copies gave rise to libraries, enabling people in all social and economic strata to gain access to ideas and information previously available only to the rich. News and information were now being democratized. In addition, and equally relevant to our story, the printing press enabled science to take giant steps forward in the 16th and 17th centuries. Historian Elizabeth Eisenstein emphasizes that it wasn't just the speed with which printed books helped to disseminate scientific findings, it also was the greater accuracy of printed documentation over error-prone handwritten records.

It's safe to say that, even 400 years ago, the printing press as a technological innovation was rendering certain jobs obsolete. But many more new jobs were being created. Though trained artisans were no longer needed to create and illuminate manuscripts, thousands of people now found work in printing shops, bookstores, libraries, and in new social settings like the coffee shops of London, where crowds of readers would gather to peruse the latest newspapers and discuss their contents. And once book publishing became an industry, "author" became a recognizable, paid occupation for the first time in history. Clearly, the impact of technological development on human prosperity has a long tail!

The paradigm shift created nearly singlehandedly by Gutenberg led to shifts in European culture and human thought that later paved the way for globalization. It catapulted all three key drivers—innovation, globalization, and demographic shifts—to new and lasting levels of influence that continue to shape our world and our worldview. As Marshall McLuhan wrote in his book *The Gutenberg Galaxy*, technologies are not simply inventions that people employ; they are also the means by which people themselves are reinvented. Ultimately, McLuhan credits the invention of the printing press with helping to spur the rise of nationalism, automation of scientific research, standardization of culture, and alienation of individuals, among other developments in the last half millennia.

Standing at the edge of continued and even greater human-technology convergence, some more recent students of media impact, like Sven Birkerts, author of *The Gutenberg Elegies,* are exploring how the electronic age now threatens the very literacy that Gutenberg's invention evolved. Drawing on his own love of books, Birkerts examines how "hearing" a book or reading one off an electronic screen diminishes the experience of reading and may eventually weaken our literary culture.

This is how the driving forces get linked. We can't look at the impact of technology without considering the supporting role it has played in the evolution of the other two driving forces.

For example, it's difficult to overstate the role technology has played—and will continue to play—in the rise of globalization. The internet has been called the backbone of globalization, enabling the real-time communications that have helped erase major barriers to collaboration. It allows a colleague in Tokyo to share projects with her counterpart in New York as easily as if they were sitting in the same room. The internet has made it possible for companies to expand their footprints and coordinate their global workforces in a follow-the-sun approach that gives clients everywhere access to 24/7 service. Meanwhile, globalization has helped create a consumer boom, ushering in unprecedented economic opportunities. And it's given consumers a veritable world of options to choose from, whether they're buying a new pair of jeans or an electric car.

Meanwhile, the convergence of globalization and technological innovation is paving the way for healthier, safer, and overall more comfortable lives for most of the world's citizens. A few examples:

- Childhood mortality rates have plummeted, as have rates of extreme poverty. In 1800, 43 percent of children born around the world died before their fifth birthdays. Today, fewer than 4 percent do.
- Over the past 50 years, global gross domestic product (GDP) has increased exponentially, growing

from just \$1.4 trillion in 1960 to more than \$85 trillion in 2018, according to the World Bank. Put even
more dramatically, just in the past 25 years, GDP
increased by about the same amount as it did in the
25,000 years preceding the year 2000.

- Also in just the past 25 years, the fraction of people
living in extreme poverty dropped from 36 percent
to less than 10 percent of the population globally.

Of course, globalization hasn't been all good news. In
some cases, the promise and potential of improved global
reach has been exploited. Instead of bringing a more humane approach to labor conditions to some emerging
economies, globalization has resulted in some companies
opportunistically seeking out markets where employees
are paid less and are treated less humanely.

Thus, there's work to be done to realize the potential
benefits of globalization. The good news is that, in many
ways, the global community has never been more unified
in addressing serious threats to humanity. Issues like climate change, safe food production, water conservation
and access, and nuclear war are global concerns, and are
receiving critical global attention.

I believe it is essential to embrace the spirit of economic and cultural integration, and to bring those forces to
bear on these critical issues. Doing so will help ensure that
the our more close-knit world will be a force for positive
changes to benefit humanity.

Breaking Down Generational Divides

AT THE SAME TIME THAT we are called upon to channel the force of globalization so that its benefits will be shared by all humanity rather than just a select few, we are also being summoned to find ways to bridge the demographic and generational divides that threaten our ability to cooperate as a species.

Each summer, the firm I work with hosts about 2,000 college interns across our U.S. locations. In August, the entire group is brought together for a few days to celebrate the successful completion of their time with us. I have had the pleasure of speaking to this group a few times in recent years. I always learn something at these events, and being with these motivated high performers keeps me fresh and on my toes.

By 2019, this group was solidly Gen Z, having been born right around 1995, plus or minus a year or two. While each generation is assigned certain characteristics by which it will forever be measured, this one seems to have more than the usual share of expectations riding on it.

Some experts predict Gen Z will act to bridge all other generations, breaking down the concept of generational thinking that has often divided us. Where we were once defined by our methods of working and our willingness to adopt new ideas, we are now unified by our technology tools and interconnectedness. Grandparents and toddlers both use iPads. Similarly, the idea is that Gen Z is relatively "age-blind," focused on the value a person can bring and the difference they can make rather than on the tools they use.

The collective goal of Gen Z is to bring us all together in networks and communities of influence. After all, this is the generation of Uber, crowdsourcing, and interactive smart devices. One has to reflect no further than to consider a few powerful examples: the students of Marjorie Stoneman Douglas High School taking up the mantle of gun control after the Parkland shootings; Gen Z leadership of the marches against racial injustice in 2020; and the tremendous success of many social media crowdfunding drives to help the ill or disadvantaged. Today, even people who are younger than the interns I met in 2019 are leaders on the world stage, with Nobel Peace laureate Malala Yousafzai (born in 1998) and environment activist Greta Thunberg (2003) as two examples.

With this as a backdrop, we viewed a recent intern conference as an opportunity to explore their thoughts on issues of the future and the skills needed to address. The results of a survey we did with them offer insights into what we all can expect:

- Seventy-six percent of the interns responded to the survey, which demonstrated immediately that this is a highly engaged group that does not shy away from sharing its views. We have rarely received such a high rate of response to any online survey.
- When asked which skills they considered most important for success, their value-adding and community-minded attitudes shone through, with relationship building and critical thinking topping a

list that included many likely options, such as intelligent automation, data analytics, and innovation.

- Hands-on experience was their preferred means of learning for career advancement, and they take personal responsibility for their career progress rather than assigning it to academic institutions or employers.
- We hit the demographic wellspring with our last question: "Select up to three issues from the list below that you believe are most likely to be top global challenges faced by your generation over the next 10 years." Sixty-five percent selected "Maximizing human capability in an increasingly technological world," followed by climate change at 58 percent and the future of education and diversity and inclusiveness tied at 41 percent.

Surveys of young people across the globe are generating similar results. The Varkey Foundation, a UK-based nonprofit focused on education, conducted a study of Gen Z participants in 20 countries across all continents in 2017. As in our survey, these global young people were tuned in to key issues around humanity and turned on to technology:

- Just over two-thirds thought that making a wider contribution to society beyond themselves and their family is important.
- Overall, young people in 16 of the 20 countries said they are pessimistic about the future, concerned

that the world is becoming a worse rather than better place to live. But 84 percent said that technical advancements make them hopeful about the future, and 80 percent had faith in the power of education as a source of hope.

- Nineteen percent said that greater skills would help them make a bigger contribution to society, and 26 percent said that more knowledge would do so.
- Overall, 68 percent said they were happy, and those in emerging economies tended to experience higher levels of happiness than those in so-called developed countries.

I think these samplings of the up-and-coming generation say a lot about just how squarely their feet are on the ground. They are involved, they are accountable, and they recognize the pragmatic and existential business, educational, and social challenges that lie ahead. Hyper-connected and democratized members of Gen Z see us all stronger as one, even as they place great value on the contributions of the individual. I believe they will make great employees, loyal friends, and solvers of big issues. They give us hope for humanity's future.

SPOTLIGHT

The Monitored Body—
The Rise of Technology-Enabled
Healthcare

Wearable and embedded technologies continue to gain popularity as two useful tools for health mapping. Such digital tools improve the timeliness and quality of care. They provide us with moment-to-moment, data-driven feedback on basic health and fitness metrics as well as the ability to monitor specific illnesses, chronic conditions, and other physical states. Typical versions of these technologies are able to monitor the user's daily activities and crucial biomedical data, ranging from glucose monitoring and blood oxygen levels to distance traveled, number of steps taken, heart rate, calories burned, female hormonal levels, weight-lifting recovery time, sleep quality and duration, and many other vital signs and statistics.

Health and fitness monitoring devices are just one small, familiar example of the way innovative technologies are beginning to transform healthcare, with implications that go far beyond the obvious short-term impacts.

With 80 percent of today's healthcare costs—and the majority of deaths each year—related to chronic, lifestyle-driven illnesses, there is a steadily growing focus on patient empowerment and engaging the individual in their own active treatment. This shift has given rise to a second healthcare trend that exploded in importance and popularity during

2020: telemedicine, the practice of one-on-one, human-involved healthcare from remote locations facilitated by electronic video and data connections.

In recent decades, a number of healthcare providers and tech companies have sought to popularize telemedicine, especially for people living in remote, often rural communities and those suffering from rare conditions that demand the attention of a specialized professional likely based in some distant location. There are obvious time, cost, efficiency, and productivity benefits to telemedicine. But, perhaps understandably, most patients and many healthcare providers were reluctant to give up the familiar, intimate connection provided by an in-person, face-to-face consultation.

That changed with COVID-19. The pandemic forced the closure of countless medical offices and made many patients anxious about the idea of sharing a small, enclosed space with a medical professional who might well have been exposed to the virus. Telemedicine became the first option for millions of healthcare consumers rather than a distant, disfavored alternative. Now many are recognizing the unexpected benefits this mode of care provides, including the speed and ease of arranging appointments, reduced travel costs, and less need to take time off from work or to manage child care.

Telemedicine also gives patients access to a greater variety of providers and new technologies. It helps address the reality that chronic conditions lead to long-term relationships between doctors and patients, lasting not just months but often years, even decades. And experience has shown providers that telemedicine-based practices incur far fewer missed and cancelled appointments.

As a result, telemedicine is expected to grow into a $66 billion global industry during 2021, according to a study by market research and advisory firm Mordor Intelligence.

As these and other new medical technologies continue to proliferate, they are helping to usher in a new way of thinking about healthcare—one that might be described as "connected holistic wellness." The traditional popular conception of healthcare views the body as a set of disconnected systems that periodically malfunction, requiring the attention of a professional "mechanic"—a doctor or other provider. By contrast, this new approach to wellness emphasizes the interconnections among all aspects of fitness and the intimate involvement of individual human beings in monitoring and maintaining their own health. Exercise, nutrition, activity levels, work and career satisfaction, family and social relationships, psychological and spiritual fulfillment—all are deeply intertwined with physical health when the holistic wellness model is embraced.

In the years to come, this new approach to healthcare is likely to combine the use of technological tools with counseling, coaching, and patient education to create a holistic wellness model customized to the needs and wishes of every user. Health-related data may be gathered from a large number of computer-interfaced sensors embedded in the individual or in objects the individual touches, wears, uses, or otherwise interacts with. One or more user-owned healthcare devices will translate the data into simple, useful information, not just reporting about the user's current fitness level but also offering guidance and advice about diet, exercise, sexual activity, travel, work, and other important matters. Biomedical data like blood sugar level, blood pressure readings, pH balance, body mass index, and hormone levels may be captured and shared with multiple practitioners involved in the individual's care, from physicians to nutritionists, trainers, and therapists.

A suite of technology tools—digital therapeutics—designed to support individuals in their quest to monitor chronic conditions and manage holistic wellness could have the potential to produce enormous healthcare benefits—longer lives, greater fitness and vitality, better functioning for those with disabilities, improved management of chronic conditions, and more.

But like all technologies, such tools can be used with greater or lesser degrees to social and ethical responsibility. Here are some of the kinds of questions that technology experts, healthcare professionals, policy makers, and concerned citizens need to be thinking about today:

- Who will own and control the data generated and gathered by the new healthcare technology tools?
- How can individual rights to privacy be protected even while data-sharing is used to provide healthcare professionals with the accurate information they need?
- How can we draw appropriate lines between policies that incentivize healthy behaviors—for example, health insurance discounts for users whose data show they have adopted positive lifestyle activities—and those that "punish" people for personal choices they should be free to make?
- How can aggregated data from the new healthcare technology tools be used in ways that benefit society (for example, by supporting research into important

new cures) rather than facilitating purely private advantages (for example, in marketing and selling costly treatments with minimal proven value)?

It won't be easy to come up with the right answers to questions like these. In some cases, technology itself may provide some of the solutions. For example, blockchain technology, which allows information to be securely stored while allowing access granted upon permission, might have potential as a tool for safeguarding the privacy of healthcare data while letting the individual decide what information to share with a chosen team of practitioners.

In the ideal future we all want, the new medical technologies should lead to treatment conditions in which the consumer is in control, both of the process itself and of the potential trust, privacy, and ethics issues surrounding it.

CHAPTER THREE

The Big Shifts: Six Major Trends Shaping Our World

GLOBALIZATION, THE PACE OF INNOVATION, and major demographic shifts have truly upended our lives. These three drivers can be considered the motors that are propelling the transformation of our fast-changing world. A closer look, however, shows a number of trends associated with or subsets of each of those drivers that are contributing to this ongoing transformation. Even a trend that appears minor, slow-moving, and short-lived can compound to generate enormous social, economic, political, environmental, or technological change. In other words, a seemingly unimportant trend can unexpectedly morph into what we might call a *megatrend*.

Here's a simple example. In 2003, MySpace was launched to little fanfare. Within a few years, the site was among the most visited on the nascent but fast-growing internet. For most, however, MySpace remained merely a curiosity—a substance-free, teen-centric corner of the Web.

Today, of course, we all know where the story goes. Facebook, which was launched in 2004 as a rival of MySpace, exploded in popularity and led to the mainstreaming of social media. These days it's hard to find anyone whose life is wholly disconnected from social media. Social media is how companies reach consumers and how politicians reach voters. It's where many of us get our news and updates from our families; it's where we share our happiest moments and bitter defeats. Thus, the birth of MySpace and the concept behind it ultimately gave rise to a major force transforming the worlds of culture, media, politics, and technology—a megatrend by any measure.

What, then, are today's megatrends? I believe there are six key megatrends that will bring their own forms of transformation to the people and institutions they touch. They are tied strongly to people's needs and cultures, and they are all shaped to some degree by technology. Unlike the driving forces discussed in the previous chapter, the effects of these megatrends are powerful but evolving, and some may prove to be relatively short-lived given today's standard of fast-paced change. But at least some of these megatrends are likely to play key roles in restructuring and redirecting our lives over the next five to 10 years.

Trend 1: The Rise of Autonomous Systems Driven by Big Data

AS WE'VE ALREADY SEEN, the past few years have seen big strides in technology designed to augment human capabilities—and that augmentation is increasingly becoming active rather than passive. No longer merely assisting people, technologies are taking an active role in helping people think, learn, and work. They're working with people to conduct sophisticated and functional tasks, from performing surgery to remotely flying an aircraft.

The data and processes that support these activities are known as *autonomous systems.* These systems span a wide range of disciplines, including data mining and analysis, neural networks, pattern recognition, and machine learning. Autonomous systems are incredibly sophisticated, capable of handling complex mathematical and scientific theories and applications that are beyond the comprehension of most technology users. But while the average person may not understand *how* autonomous systems work, they know what these systems make possible. Autonomous systems are the technology that makes self-driving vehicles possible and lets commercial drones deliver their packages with astonishing accuracy. They put the brains in our smart thermostats and make our smart speakers so much cooler—and infinitely more informed and responsive—than any traditional home stereo system.

To be truly autonomous, a system must be able to do three things: gather the data required to address a problem, analyze the data to arrive at a solution, and take action to implement that solution. Self-driving cars put this into practice: The vehicle's sensors gather data. Then, intelligent software analyzes the information and makes decisions about next steps. Finally, a programmed sequence of actions executes those decisions, whether by braking suddenly in a traffic jam or taking a left turn on a busy city street. These systems practice autonomy—the ability to act independently of direct human control—and often do so in circumstances that are uncontrolled and therefore more or less unpredictable.

Autonomous systems are one of the top megatrends we humans need to monitor. The next wave of highly disruptive technologies is likely to be fueled by such systems, including wearable technologies, virtual reality, blockchain, and the continued evolution of self-driving vehicles. Autonomous system developments are behind greater automation in the workplace, whether it's on the manufacturing floor or in the C-suite. They will enable breakthrough discoveries in health and wellness and reshape financial and professional services.

Autonomous systems are also likely to further blur the lines between humans and machines. Here is where the concept of the robot comes in.

Karel Capek, a Czech intellectual and writer, was the first to use the term *robot* in his play *R.U.R.,* produced in Prague in 1920. Capek's storyline, built around the fictitious corporation known as Rossum's Universal Robots, featured man-made, human-like machines being treated as slaves and then eventually turning on their creators. His point was not to elevate robotic capabilities but to underscore the threat of machines adopting or even exceeding the power of the human brain. "The product of the human brain has escaped the control of human hands," Capek said in an interview after the play opened.

For most of the past century, the fictional idea of robot capabilities as described in books and movies (from the classic 1927 film *Metropolis* to *The Stepford Wives, Terminator,* and *Blade Runner*) outpaced engineering reality.

Today, however, actual robotic capabilities are well out in front of many of their fictional ancestors, spurred on by developments in artificial intelligence.

One of the first robotic devices employed in real-world work was a robot arm installed at an automobile factory in 1961 to assemble, weld, and paint vehicles. Today such machines are commonplace. The *Wall Street Journal* reported that 373,000 industrial robots were sold in 2019, for a total of 2.7 million in use globally. Most of them work hard but are not very smart, fulfilling similar tasks to those performed by their 1961 ancestor.

The newest robots, however, are equipped with sophisticated software and hardware tools that make them capable of managing warehouse inventory and order fulfillment, disinfecting hospitals, making deliveries, and assisting with telemedicine applications. As discussed in chapter one, software robots are also handling a lot of predictable and repetitive work in offices. Thus, robots are making the transition from displacing people in manufacturing to replacing increasing numbers in the service sector.

One of the big breakthroughs that is empowering the latest generation of robots was the development of Lidar, a light detection system that creates a 3D map of a physical space using laser beams. Lidar is a key to the development of self-driving cars and unaccompanied robotic delivery vehicles. It's often combined with multidimensional cameras that allow a robot to use algorithms to pick out landmarks and identify objects, which helps to keep them from running into things. Newer technologies currently in development will soon allow robots to "see around corners" by picking up on subtle movements of light and air that the human eye can't perceive.

Beyond handling more intelligent tasks and avoiding objects in their path, another long-time objective has been to make robots more graceful. Whereas biped robot models were formerly notable mainly for their awkward joint movements, more recent models equipped with actuators (motors inside robotic joints) can carry out heretofore impossible sets of motions such as doing pushups and backflips.

Robots are also getting smarter. Thanks to AI programming, they are increasingly capable of combining knowledge with sensory processes. For example, they can sense the heat from a hot machine and make a smart decision about whether and how to touch it. In a lab at UC Berkeley, a humanoid robot named Brett has learned how to solve a puzzle made for preschool children where pegs of different shapes are put into matching holes.

Some specially-designed robots are performing a humanistic purpose. These are animal-mimicking robots that serve as electronic pets for the elderly and lonely— machine companions filling a very human void. Their animation capabilities include blinking and closing their eyes, purring, barking, squirming, yawning, and curling up in a lap. Owners of these animatronic dogs and cats gain companionship and something to pet, hold, and even to give meaning and orientation to their daily routines.

Providing such robots to those who need them is not just a nice thing to do. Medical research shows that social isolation and loneliness can inflict a lot of harm on the human body. Research by experts at Stanford University and the AARP (formerly the American Association of Retired Persons) has found that social isolation adds nearly $7 billion per year to the cost of Medicare, because lonely people show up at the hospital sicker and stay longer.

For all their growing capabilities, robots remain a tool in need of a lot of guidance. In general, robots are not yet capable of performing in an unstructured environment without human support; a human still has to push the button.

In the near future, however, this will change. Systems far more advanced than those found in Capek's robots will begin making decisions at higher levels. When that happens, how can humans ensure they will still have a hand on the steering wheel of these technologies?

Doing so may require a fundamental reframing of our relationship with technology and autonomous system design. Businesses will need to recognize the awesome power of these tools and the responsibility that comes with their use. Governments likely will have a critical role to play in regulating autonomous systems to ensure that society can enjoy the benefits of these powerful technologies while public safety, data privacy, and fairness are all protected. Some of the problems that must be addressed are already rearing their heads. For example, some autonomous systems have been shown to have racial, gender, and age biases, as well as other prejudices, unintentionally built into their critical algorithms.

The fact is that the evolution of autonomous systems is tremendously promising, but it also gives rise to numerous difficult questions around the meaning of autonomy, identity, and privacy. How independent do we want autonomous systems to be? Up to what point can a system think and act independently as it grows and develops? Who is responsible for its actions or inactions as they impact society? Ultimately, the most important question to consider is how we'll decide when the benefits outweigh the risks.

Trend 2: The Empowerment of Consumers

CONSUMERS ARE ENGINES of the world's major economies. In the U.S., consumer spending accounts for a whopping 70 percent of GDP. So when there's a fundamental shift in the relationship between consumers and the companies that serve them, the impact on the economy as a whole is profound.

That's exactly what has happened in the last couple of decades. Thanks to the combined impact of a host of specific changes—the rise of global business, the advent of the internet, improved methods of communication and transportation, the steady increase in the number of people around the world who can afford a middle-class or affluent lifestyle, and other such trends—consumers have become far more powerful than ever before.

Increasingly, consumers are in the driver's seat in their relationships with companies. And they want more than just great products and great prices. Eighty-four percent of consumers say that their *experience* with a company is just as important as the products or services it offers. And with so many manufacturers, retailers, and service companies vying for their business, why should consumers wait to sat-isfy their need for tailored experiences?

Consumer-facing companies are now facing a buyer's market in every sense of the term. It's no longer enough to design and manufacture good products or to provide useful services; companies today must leverage data to deliver highly personalized experiences and customized products that are precisely what customers want, when, where, and how they want them.

Technology is stepping up to satisfy these new de-mands. Virtual personal assistants take and place orders; same-day messengers and drones deliver packaged items. Taking these concepts even farther, the world's largest online retailer applied for two patents in recent years that may enable it to continue expanding its huge base of loyal customers for some time to come. One, a patent for what it calls "anticipatory shipping," allows the company to ful-fill an order before it is even placed. Using its massive da-tabase of shopping histories, the company anticipates when a reorder is due and ensures the package is available at the shipping hub closest to the consumer.

The second patent is for offering 3D printing of prod-ucts in its delivery trucks. With this one, it seems everyone wins—the company saves money on inventory control and warehousing, and the consumer saves time on delivery. One-day delivery, which the company is already known for, nearly accomplishes the same result as on-site "manu-facturing," but the infrastructure and tools required for fast delivery of a stored item are an added cost over delivery van production.

This type of thinking and planning—a commitment to seeing around corners for where the next opportunity might be—is drawing a clear line between manufacturers, wholesalers, and retailers who succeed in the new era of consumerism and those who don't. Collecting and crunching data, anticipating and embracing change, investing intelligently, and continually innovating—these are among the skillsets that business leaders must master and practice in the twenty-first century to keep up with the changing needs of consumers.

Even small shifts by the American consumer can have a profound competitive and economic impact, but developments like these are anything but small shifts.

Technology itself is one of the elements driving consumers' higher expectations. The ability to surf a world of retailers from the comfort of your couch is absolutely empowering buyers. And technology—data and analytic tools, in particular—is a key ingredient in how companies are attending to consumers' needs. But there are other factors contributing to this major shift in the consumer landscape, including changing generational attitudes. In survey after survey, members of Gen Z clearly express their disdain for all things on-the-rack and off-the-shelf. Whether they're shopping for clothing, food, financial advice, or healthcare, Gen Z and Millennials are today's economic power brokers: They expect exactly what they want, when they want it, and at the best possible price.

As the science of predictive analytics continues to improve, companies will be able to use data more effectively to get out in front of the next big consumer trend. Investing in new, disruptive technologies driven by predictive analytics will be essential for survival. Those investments will also enable companies to develop the business models, marketing strategies, and supply chains that let them deliver the kinds of tailored solutions—and gratifying experiences—that meet consumers' increasingly lofty expectations.

The good news for business is that these shifts tend to open the spigot of innovation, and the ideas that emerge can give entrepreneurs and startups the ability to compete with even the most entrenched players in the market. In recent years, we've seen a burst of innovation as companies have had to figure out how to serve the new consumer. It's led to creative ways to seamlessly deliver goods and services. It has also fueled surprising and pleasing experiences for customers, both digitally and in real life.

The availability of data via smartphones, social media, search engines, purchase histories, and more has given us what some call *the behavioral economy*—one in which data-generated intelligence is used to appeal to and influence consumer decision-making.

This doesn't mean that data-savvy companies will find it easy to figure out exactly what customers want. Being truly inventive is about far more than giving customers what they think they want. As Henry Ford purportedly said, if he had asked customers in the era before automobiles what they wanted, they would have said "faster horses." Today's consumers want more than responsive product recommendations. They also want smooth, frictionless transactions, rich experiences, and relationships built on trust—and the definitions of these desiderata continue to evolve and become more demanding over time.

Consider the example of financial services and financial technology (fintech). It's an industry caught up in appealing to at least three generations, each with quite different expectations. Some baby boomers still use checkbooks and like going to a physical location where they can talk face-to-face with their trusted personal banker. Millennials, the largest generation in history and likely to soon control the largest portion of the world's wealth, seek innovations like mobile banking and access to new financial investment instruments like cryptocurrencies. Gen Z, which treasures a sense of belonging and purpose, may be attracted to more low-tech, high-touch strategies—for example, the availability of personalized financial strategies designed to fit their unique circumstances.

Meanwhile, many consumers, especially younger ones, are beginning to abandon banks altogether. A report by the World Economic Forum predicts that the traditional banking model will gradually be "unbundled." As trust in non-bank fintech providers grows, they will get a steadily larger share of the customer wallet, from credit cards and mortgages to investment management. Users of one very popular social media app in China, for instance, need only click a single icon to exchange messages, book a car service, order dinner, schedule a manicure—or arrange a money transfer. In a world where that is possible, many ask, "Why on Earth do I need a bank account?"

Regardless of the industry, AI technologies will support disruptors and innovators who see new, unrealized opportunities for consumers in the treasure troves of data that are constantly being collected about their needs, desires, and preferences. Evolving AI tools will also help companies boost revenues by enabling automation efficiencies, reduced errors, and better resource allocation. The next step, according to some AI consultants, is the ability to tailor products to individual preferences—a high-tech update on the old-fashioned concept of "bespoke" products made by a craftsman using hand tools.

Human augmentation technologies, including virtual reality tools, will also play a role in customizing consumer experiences. Imagine an avatar that replaces today's in-store salesperson and online virtual associate, speaking with you and walking you through a set of options tailored to your shopping history and needs. When you're shopping for clothing, your sales avatar will know your measurements and can recommend styles you will like. Industry leaders will value and use their access to personal information wisely, empowering their customers rather than exploiting them. In this time of diminishing trust across all institutions, this approach can help build and bolster brands.

Consumers are also looking for brands that reflect their values—and that's becoming more challenging over time, as each generation has very different wants and needs. It's not enough to offer products a consumer wants; engaging customers may also require companies to define and articulate a particular point of view or sense of purpose. Take Gen Z: this generation wants to feel part of causes they can help to create and support, and they will look to companies that reflect their values in everything they do.

Finally, companies need to reckon with fundamental challenges to the very concept of consumerism. Global issues like climate change and trends including aesthetic minimalism, smaller living spaces, and ride- and home-sharing services are driving a distinctly anti-consumption movement. It's a movement that prioritizes access over ownership: Use what you need, enjoy the experience, then just walk away. Indeed, we may find that the most empowered consumers of the early 21st century will have gained a satisfying experience while barely "consuming" anything at all.

Trend 3: The Transformation of Healthcare

HEALTHCARE IS THE MOST PERSONAL of all industry sectors. It touches people directly and indiscriminately, impacting body, mind, and spirit. It can also have a ripple effect, impacting families and social communities. Ironically, it is some very nonhuman bits of technology that are now disrupting this very human endeavor. Metallic threads, chips, and screws are contributing to cures and treatments that even a few years ago would have been considered miraculous. Yet these high-tech advancements are putting the human—the individual consumer—more firmly in control of the future of healthcare.

The concept of "personal wellness"—the notion that individuals are responsible for their own health and need to be actively engaged in monitoring and improving it on a daily basis—is a relatively new force in the world of healthcare, and one that is growing steadily in importance. Technology is empowering people to take charge of their health and well-being in some very practical ways, such as monitoring and enhancing sleep patterns, helping manage stress, and enabling better tracking of fitness activities and their benefits. Individuals are equipped today with more data that allows them to make proactive lifestyle choices and changes, which also gives them more control over their health and decreases dependency on healthcare professionals.

The human body has more than 35 trillion cells, producing billions of chemical reactions every minute. That is a complex system at work, but it also is a treasure trove of useful data. Technological developments in the life sciences will enable equipment and algorithms to look for patterns and insights, enhance understanding, and facilitate connections that currently are missed. Meanwhile, greater mobility, cloud systems, and digital diagnostics are all contributing to make healthcare more accessible and effective. Treatments are becoming more precise and targeted. Access to healthcare is exploding thanks to virtual visits, wearable diagnostic tools, and electronic records.

Then there are the critical questions about privacy, ethics, and trust. With all these technologies and digital tools, terabytes of data are being collected every second about our most intimate details. What happens if our data are used for experimentation, exploited for commercial gain, or released into the wrong hands? Should we be more concerned about the ethical considerations emerging from the convergence of technology and our bodies? How much do we *really* trust technology to manage our health? At what point will we be so thoroughly augmented by technology that we are no longer really human?

Data and privacy challenges, then, represent major healthcare issues we need to address. Others include:

- *Growing costs and shrinking funds.* Despite standing on the brink of huge technological breakthroughs, healthcare institutions often lack the funds needed to support transformative efforts.
- *Unequal access.* As the number of physicians declines and a large percentage of the population continues to live in rural areas, access to advice and treatment remains an issue, even considering recent tech-related developments like telemedicine.
- *Uncertainty regarding government support for healthcare.* In the U.S., an aging population, growing pressures on infrastructure, and the need for more investment in innovation is making it difficult for governmental leaders, regulators, and lawmakers to find ways to address critical healthcare challenges.
- *The demand for universal healthcare.* In the wake of the COVID-19 pandemic, the issue of universal healthcare coverage may receive increased attention. As companies recover from the lockdowns of 2020-21, they are likely to employ more gig workers or self-employed contractors. There are also likely to be more of both of those types of workers in the labor force due to loss of full-time jobs in 2020-21. This trend will put more pressure than

ever on lawmakers to protect the healthcare access of the vulnerable, including both part-time or self-employed workers and the unemployed.

Without a doubt, healthcare's promise as an industry and as a support structure for human well-being is vast and encouraging. The continued merging of technology and healthcare mirrors the convergence of technology and humanity. These unique marriages can help make us better, stronger, and more fully alive. They can solve some of our most fundamental and basic human fears. To reap the full benefits of the new healthcare, however, we must carefully consider how to manage the challenges and resist the temptation to over-engineer our most intimate relationship—the one we have with our own bodies.

Trend 4: The Transformation of Work

FOR GENERATIONS, technological advances have been changing how and where we work. But in the last 20 years, the pace of change has accelerated faster than at any other point in history. As we've already observed, in the last few years, robotics has moved beyond the manufacturing floor and into white-collar cubicles, using artificial intelligence and machine learning to automate time-consuming tasks that used to require human oversight. Most recently, the COVID-19 pandemic has motivated big shifts toward remote work and virtual communications, forcing many organizations to rethink their workforce models—full-time, part-time, gig, and contract, the numbers of employees they retain, and the ways they use technology to connect to one another as well as to customers, clients, and other stakeholders. In short, the pandemic has opened doors on new ideas and forced adaptations that will alter our decisions and work behaviors long after the health crisis itself is well managed or defeated.

Some of these changes were coming to the workplace anyway, but pandemic-driven necessity has accelerated their pace. Furthermore, the results many companies achieved during the pandemic showed that not only could we accomplish more using digital channels and working remotely, in some instances we could do it better, more efficiently, with less personal stress, and with greater organizational productivity. To sustain the positive outcomes and mitigate the downsides, it is important that business leaders act quickly to provide workers with clear guidance about what the future holds and with the skills and tools needed to perform well under the new and emerging norms. These norms could include greater flexibility about where and when one works, less time spent traveling and commuting, and a break for our environment in terms of reductions in carbon footprint.

These same changes might also yield new opportunities for workers to learn and grow. For example, more workers may be able to benefit from training programs, networking opportunities, and mentoring relationships thanks to virtual meeting tools, which reduce or eliminate such problems as travel costs and days away from the office. More flexible work rules may also enable employees to devote more of their time and energy to higher-level activities like strategy development and planning.

Efforts to bring all of these trends together for the best positive outcome should start with putting people first, including both customers and employees. Understanding what people really need and want isn't always easy. For example, workers want some flexibility, but they like a certain amount of structure, too. It's great not to have to do too much business travel, but coming together in person is sometimes the best option. And different cultural propensities and personality types must be taken into account, too. Leaders should keep a steady focus on ensuring balance, respect, and opportunity from a diversity and inclusiveness perspective. Companies will need to demonstrate their ability to pivot flexibly with the times, while also acting with thoughtful intentionality.

As technology continues to weave itself into our working lives, it's important for us to carefully consider which roles humans should play and which should be left to technology. We must explore how we'll collaborate and coexist with our mechanized colleagues, not just for the sake of efficiency but for the benefit of humanity, society, and economic progress. To that end, what are the basics for adapting and succeeding in this new landscape? What new roles are evolving? How should we educate and train workers to stay relevant and productive?

When a new technology disrupts the processes that people perform, we often end up with welcome changes that take the robot out of the human. This leaves people to focus on more important work—on tasks that require deeper analysis, higher-level critical thinking, and human relationships. Many hope it also will inspire a transition to more balanced lifestyles.

Working together, humans and smart machines are an unstoppable combination. Don't allow your organization to become a victim of today's era of unprecedented change. Instead, seize on this time to be innovators, empowered thinkers, and better-performing teams at whatever you make or do.

The changes now sweeping the world of work demand a new level of agility on the part of organizations. Agility is what paves the way toward building authentic relationships and the courage to explore and try new solutions. Agility helps us develop the intuition to see around corners, steer through the fog, and articulate future directions that inspire cooperation and collaboration. That's why there are many who believe our agility quotient (AQ) and our emotional quotient (EQ) are more important predictors of success than our IQ.

Keeping up in a world undergoing constant change requires sustained organizational agility in these forms:

- Faster, flatter, more flexible teams
- Decentralized decision-making
- Empowered individuals
- Creative, collaborative, and innovative teams
- The ability to reconfigure processes and relationships in real time

In addition, these three specific requirements will drive future success:

- The knowledge, skills, and ability needed to work in a digital environment

- A commitment to constant improvement, continuous learning, and ongoing adaptation
- An understanding of the need to add value beyond what technology can provide

In this changing world of work, it's also crucial for business leaders to understand the business landscape more deeply than ever before. A sophisticated, up-to-date knowledge of marketplace issues, including those specific to your industry and the industries of your customers, will help you stay competitive. Find inspiration in the change you see in other organizations, and keep learning and helping your employees build the skills they need today and tomorrow. Awareness of new trends across a broad spectrum of business and society will better prepare you to anticipate important changes and to be more resilient and responsive when change occurs.

The future of work is really about the future of humanity. Leaders must be transparent with employees and work with them to foster skill sets that will continue to be relevant, no matter what changes the future may bring. In particular, as members of Gen Z become a larger part of the workforce, we must appeal to their specific generational characteristics and their need for collaboration and genuine engagement. Whatever your size or industry, it is crucial that management work in tandem with employees to plan the way forward.

The working world has changed in ways that challenge our assumptions about what it means to be a leader. Business acumen, authenticity, talent building, and a focus on customers and results all are still extremely important, but those skills need to evolve. For example, leaders can't be comfortable just managing change. They need to be eager to embrace disruption and innovation. Most important, they need to ask what they can do now to better prepare their employees for future success.

Trend 5: The Rise of Corporate Social Responsibility

LEADERS OF BUSINESS ORGANIZATIONS can no longer stay on the sidelines when it comes to addressing the most pressing issues of our time, such as climate change, social justice, and income inequality.

An EY study in 2019 surveyed hundreds of business leaders at the world's largest companies—CEOs, board directors, investors—to better understand how they view their roles in solving humanity's greatest challenges. The survey results showed that today's business leaders as well as the investors who support them increasingly believe that their action on these key issues is directly linked to their companies' trustworthiness and ability to grow. A few key findings from the survey:

- Senior leaders must back their words with actions and authentically represent the company's stand on global challenges.
- Corporate purpose must be linked to the biggest global challenges, and internal transformation of a company's culture and operations must reflect that purpose.

- Organizational incentives, governance, and performance metrics must link to progress on global challenges and drive meaningful actions to address those challenges.

These findings signal a tipping point in how businesses and their stakeholders view the need for immediate corporate action. Doing the right thing through trust, purpose, and transparency has become a crucial driver of growth, talent recruitment and retention, and brand leadership. Those actions must also be present in a company's approach to technology; the shift from technology as a tool to technology as a true partner means it's more important than ever for companies to prioritize building trusting relationships between technology and humans.

Out of the movement over recent decades to hold businesses more socially accountable has grown a type of self-regulation referred to as corporate social responsibility (CSR). Main objectives or causes may vary from giving back to communities to social justice to environmental sustainability. But the bottom line is that CSR policies and activities impact stakeholder perceptions of the corporate brand, influencing recruiting and retention, customer loyalty, and investment decisions.

A comprehensive study on CSR by Cone Communications in 2017 reported that the stakes continue to get higher, noting, "Companies must now share not only what they are doing, but what they believe in." CSR, the report said, will impact a wide range of business operations, from supply chain transparency to water conservation. Furthermore, companies are now expected to be forces for positive change in society. For example, 63 percent of Americans surveyed said they hoped business will take the lead in driving social and environmental change.

Younger generations have even higher expectations. A 2019 study by Cone of Gen Z attitudes found that 90 percent believe companies must act to support social and environmental issues, and 75 percent of them conduct research to check whether a company is honest about its stand on issues.

Based on this type of research and the hundreds of conversations I've had with Millennials and Gen Z students and professionals over recent years, I believe that brands that hope to lead their industry and continue to attract and retain the best workforce over the coming decade must carry out five main responsibilities:

- Living up to a clearly stated purpose that reflects the needs of our broader society
- Putting people first, including employees, customers, and other stakeholders
- Providing employees with access to the latest technologies and the training required to keep up
- Caring for the long-term health of our planet
- Meeting these social needs while producing the revenues and profits needed to remain viable and resilient

Does this list of new corporate responsibilities make the work of business leaders even more challenging than in the past? Yes. But living in an era of dramatic change also creates remarkable new opportunities—which means that the difficulties we are called to address bring with them enormous potential rewards.

Trend 6: The Emergence of Business Ecosystems

IT USED TO BE THAT COMPANIES were largely responsible for their own success or failure. In the pre-digital, pre-internet era, companies competed on quality and on the level of resources and expertise they brought to their corner of the market. Each company had its own distinctive culture and brand, and companies operated, in effect, like islands.

Today, however, the competitive landscape has been disrupted and reshaped. The boundaries that once defined companies, industries, and sectors have largely dissolved. Digital developments aren't the only reason for this fundamental alteration of the competitive landscape. In increasingly competitive times, companies are doing what they need to do to pursue the path of least resistance, decreasing friction in their processes and reducing costs. Factors such as geopolitical shifts and changes in economic and trade policies also have played a significant role. And while these shifts have broadened the playing field, it's also become more difficult to spot new challenges and identify the emerging trends that can now drive a company's success.

As a result, a growing number of companies are blending their core sectors with other related sectors or even expanding into other industries entirely. The world's largest online retailer, for example, has moved into food service with the purchase and integration of a major health food chain and a British food delivery company. Its streaming services, initially designed as a delivery service for movies and TV shows, have evolved into a major production business, actually creating the movies and TV shows they stream. What's more, while most of its revenue comes from retail, subscription, and web services, most of its profit comes from selling cloud services to business organizations. It's a remarkable example of how the era of "one company, one product line" has given way to an era of business ecosystems in which many markets, technologies, and organizational structures interact in complex, counterintuitive ways.

Similarly, one of the world's largest makers of smartphones, tablets, and computers has jumped headlong into further diversification, introducing its own streaming media service, and more recently teasing the possibility of entering the electric vehicle manufacturing industry.

On another front, many other companies have realized the benefits of forming alliances with other companies—even competitors—to complement or enhance their processes and operations. Professional and financial services firms are also joining in these new approaches as their core services increasingly depend on technology innovations to meet client needs. Strategic partnerships can give companies a relatively low-risk way to expand their reach and target new markets. They can also help companies to satisfy the most restrictive laws or deliver the most cutting-edge technologies by combining the capabilities of a number of organizations in pursuit of a shared business goal.

Thus, in many ways, doing business today means depending on others for the success of your own company. Another way to think about it is that companies aren't alone on their own islands—they're increasingly part of a living, breathing ecosystem.

The auto industry offers a prime example of how this approach benefits both individual companies and the ecosystem overall. The coming together of auto manufacturing, energy, technology, and consumer product companies is helping to reimagine the car of the future. But this ecosystem is also changing ideas about how people and goods will be moved in the future. Without technology partners, car makers likely wouldn't be as far along on the road toward autonomous vehicles. And without car makers, consumer electronics companies would be missing out on a huge market for their products. The emergence of this new ecosystem has far-ranging impacts. It's changing traditional industry categories, business models, supply chains, and marketing strategies.

Here's a question to consider: Is one of the world's largest brick-and-mortar consumer products companies a retailer, a technology company, a transportation and logistics company, or a supermarket? In each case, the answer is a resounding yes. And the company got there by imagining itself not as an island but as an integral part of a complex, ever-changing ecosystem. As we move into the future, competition will be more about competing ecosystems than about individual companies battling for supremacy over other companies.

As companies engage in these ecosystems, they must ask a series of critical questions:

- *Where do you see your future competitors coming from?* Which industries, which countries, and which economic sectors (public or private) will challenge your company's success?
- *What other products and services could you offer your existing customers or use to attract new ones?* What credentials do you need to expand into other areas, and how do you get them? Do you need new people, new training for existing people, new infrastructure, or new branding?
- *What other organizations would make strong collaborative partners for your company?* Which links in your value chain are ripe for acquiring or merging with others?
- *What types of joint operations could you engage in to help build a better working world?*
- *What other considerations are involved in reinventing your organization to be viable over the next five to 50 years?*

There's an old saying that is sometimes described as a traditional curse: "May you live in interesting times." There's no doubt that all of us today are living in very interesting times. But with the right attitude—beginning with asking the right questions—we can transform a potential

curse into a real blessing, making the potential of interesting times into a springboard for amazing achievement and
widely shared human betterment.

Urban Farming—
A New Way to Feed
the City Dwellers of Tomorrow

Growth in urban populations over the past two decades has fueled increased interest in the concept of urban farming. It's a trend being supported by many nonprofit organizations committed to improving nutrition and protecting the environment. It's also being driven by for-profit businesses that are developing technologies adaptable to urban farming.

One example: Freight Farms, a Boston-based agricultural-technology company that has installed more than 200 farms around the world, using vertical farming techniques and hydroponic nutrient delivery systems that they say allow them to produce a yield equivalent to acres of traditional farmland inside a standard 40-foot shipping container.

Yet urban farming is an idea that, at first blush, might seem paradoxical—after all, isn't city living all about high-density, high-value development that drives the price of urban real estate sky-high?

Not necessarily. That familiar image of the urban environment still holds true in some places, as the prices of apartments and office spaces in midtown Manhattan or downtown San Francisco will attest. But many cities in the United States and elsewhere in the developed world contain significant amounts of space that are currently underutilized, including vacant lots and rooftops. In some cases, these "urban frontier" spaces have been created by shifts in economic growth from city centers to suburbs and "ring cities." Detroit, for example, became home to about 1.8 million people in the 1950s, the heyday of the U.S. auto industry. Today, with American carmakers no longer dominating the industry and with much U.S. auto manufacturing having moved to lower-cost areas like the South, Detroit's population has shrunk to under 700,000. As a result, the city contains somewhere between 20 and 40 square miles of empty space (precise estimates vary) looking for worthwhile new uses.

Faded, deindustrialized cities like Detroit aren't the only places where urban space may soon become available at reasonable cost. One side effect of the COVID-19 pandemic has been the discovery by many companies that large portions of their workforce can operate productively from their homes. Some businesses are responding by rethinking their traditional office management policies. As corporations reduce the footprints of their white-collar working spaces, once high-priced areas in urban downtowns may no longer be needed for offices, making them available for new uses. Those could include more affordable housing, parks and green spaces— and perhaps urban farms.

Another vast amount of urban space that may come to outlive its original purpose is parking lots and garages. As driverless cars become available and increasingly popular, it's possible that some long-standing industry sectors associated with traditional human-controlled automobile use will undergo game-changing disruption. Urban parking facilities are a prime example.

Consider Los Angeles County, one of the most car-dependent areas in the United States. It's estimated that the total land mass dedicated to public parking in the county is more than 100 square miles—nine times the total land mass of Manhattan. What will happen to all that space when autonomous vehicles (AVs) become the norm? Once AVs are capable of dropping off passengers and picking them up again later without the need to park nearby, the need for vast parking areas close to workspaces will shrink dramatically. Designated AV parking could be back at home, or in a remote lot some distance away, from which the car can be simply summoned on command—a kind of "valet parking," but without the need for a valet. What's more, when AVs are parked, they may take up much less space and have a much higher parking density than is possible with human operators. Thus, greater use of AVs and the more efficient parking solutions they make possible will result in a significant reduction in the number of public parking garages and lots needed.

The valuable real estate freed up could be used in ways that benefit humanity-first principles. One way could be by providing "horticultural therapy" to urban environments. Farm spaces in cities would produce measurable environmental benefits, including lower levels of greenhouse gases, reduction in the use of pesticides and other harmful chemicals, and improved air quality. Today, conventional farming uses 70 percent of the world's drinkable water, and ever-growing amounts of water are being polluted by fertilizers and pesticides. Water reuse would be an important advantage of vertical farms.

Urban farms would also create other benefits for city dwellers by creating jobs and increasing the availability and affordability of fresh produce. A 2014 article in *The Economic and Social Review* calculated that urban agriculture could reduce the typical distance that fresh food travels from farm to table from about 1,300 miles to just 30 miles, sharply reducing the amount of energy needed for transport as well as the pollution generated. Spoilage and food waste would likely also be reduced, and locally collected food scraps could be used as compost, making the city farms even more productive—a constructive, self-reinforcing cycle of positive developments.

If 200 square miles of city land were converted into vertical farms, urban farmers could produce enough food for 20 million people—fresh, organic products, locally grown and locally harvested.

Vertical farming isn't the only evolving technology that could fuel the growth of city-based agriculture. Two other examples are materials recycling and so-called additive manufacturing, the latter often denoted by its signature technique, 3D printing. Concrete and other building materials recovered from demolished parking garages, factories, and other obsolete urban structures could be repurposed for use in creating urban farming facilities. And 3D printing can be deployed to drive down the cost and complexity of such projects even further. Though many people still think of 3D printing as being restricted to small plastic objects, it is already being used to manufacture large-scale structures like houses. As of early 2021, hundreds of construction projects have been successfully completed using 3D printing, showing that a small, 3D-printed home can be built for less than $4,000 and in fewer than 24 hours. Adapting the same technology for urban farming purposes would be relatively simple.

Among the ethical and social issues we'll want to consider as we explore the possibility of large-scale urban farming are the following:

- How can abandoned or underutilized structures and plots of land in American cities be economically converted into possible sites for urban farming while providing reasonable compensation to current property owners?
- How can the costs of property conversion—for example, of soil decontamination in spaces where former industrial uses have left pollution behind—be covered without unduly burdening the business prospects of the new urban farms?
- How can we engage local citizens and community groups in planning, running, and staffing the new agricultural businesses that could spring up in cities across America, thereby encouraging inner-city entrepreneurship and ensuring that the interests of ordinary people will be served by this new model of food production?

The concept of taking advantage of such disparate technological developments as vertical farming, autonomous vehicles, and 3D printing to create positive, innovative solutions to some of today's most vexing social and economic problems represents the potential for human-technology convergence at its best.

CHAPTER FOUR

Building Our Better Future:
How Each of Us Can Respond
to the Challenges of Tomorrow

I BELIEVE THERE IS MUCH to be said for enabling technology to complement and augment human capabilities. But I also believe that humans will always hold a comparative advantage over machines in key aspects of life and business. Thus, it's important to consider which roles humans will play and which will be relegated to technology. We also must continue to explore how we'll share the work and collaborate with our evolving machine colleagues to ensure that together we reap the greatest benefits for economic progress, humanity, and society.

To that end, what are the basics required of humans for adapting and succeeding in this new landscape? What are the enduring requirements of human success? What will be the future-oriented roles and labor divisions that will define the jobs of tomorrow? And how should we humans educate and train ourselves to stay relevant and productive?

Let me start by observing that I don't believe that human work will disappear or become less of a social and economic imperative. Work is important to people in many ways, of which earning money is only one. Work is also about community and family structure, confidence, self-worth, creativity, and pride.

Jim Clifton, former CEO of the Gallup Corporation, wrote a book titled *The Coming Jobs War* after his organization conducted a major survey of global citizens regarding the future of work. Clifton observed that the driving force that motivates most people is not freedom, peace, raising a family, religion, or property ownership. "The will of the world," he wrote, "is first and foremost to have a good job. Everything else comes after that." Gallup's study showed that people find mastery and purpose in work, with purpose being especially important to the rising generations who will soon dominate the world's workplaces.

As I'll discuss in detail in chapter five, I believe the current and future direction of effective technology transformation puts a heavy focus on the qualifications of the people who will use new technology tools and their output most effectively. In these early years of the 21st century, the lens through which we view technological change needs to be recalibrated. Rather than centering our focus on which products, platforms, and algorithms make AI and intelligent automation smarter and faster, our focus should be on equipping people with the skillsets and attributes they need to use and leverage those technologies effectively while also feeling accomplished in their own right.

The other leading forces we've considered in this book only underscore the same point. Globalization and the demographic and generational transformation of society are putting a lot of pressure on the need for greater connectivity and mobility, more speed and efficiency, and innovative business models and strategies. These demands are leading to the evolution of new business and workforce models, business strategies that view disruption as an opportunity rather than an existential threat.

For all these reasons, we are living in challenging times for working men and women. In this chapter, I'll offer my thoughts on what each of us needs to do to respond to the changes that the world is throwing at us, today and in the decades to come.

The Technology Paradox: How Smart Machines Make Human Qualities More Valuable

IN TODAY'S WORLD of ever-accelerating innovation, everyone needs to be thinking about what it will take to survive and thrive in a changing future world. You may be concerned about this issue because you have school-age children and you're looking for the best advice to guide them in their studies or their search for employment opportunities. You yourself may be the student, or the teacher, or the prospective employer. Or you may be a mid-career professional wondering how the prospects for the second half of your life will compare to what you've

experienced so far. Will the same talents that have enabled you to survive to this point continue to serve you well in the world of tomorrow?

Everyone wants to know what skills, competencies, and personal attributes will have the most value in the years to come. Unfortunately, the details of the future are essentially unknowable. We're left to conjecture and project based on what we know now and on past experience, including recent trends. Dealing with uncertainty in a constructive and positive way is a challenge we all face as individuals.

In my view, the most productive assumption—as well as the most likely one—is that life will continue to require a combination of capabilities that use human advantage to its fullest in developing and exploiting technology while also strengthening areas of uniquely human advantage over technology's current limitations.

Technology is both a prime disruptor of today's business environment and a solution to the challenges it poses. New digital tools help people connect at a time when many worry that reliance on technology is precisely what is driving us apart. They enable new types of workforces—virtual, digital (bots), and offshore—which contribute to our efficiency. However, the digital tools are not yet supported by enough people with the right skills to help them and us reach full potential.

While machines get better all the time at mirroring or mimicking human ability—for example, by beating chess grandmasters and Jeopardy champions at games of pattern recognition and fact recollection—they remain, at least for now, less threatening in areas that require ideation and the

ability to "see around corners," to feel and express empathy, to self-produce complex forms of communications, and even to apply basic human senses such as sight, smell, and touch. Thus, if you want to define what makes human beings uniquely valuable, think about jobs that require one to ask good questions, not merely to generate accurate answers, as well as work that demands curiosity, creativity, thinking ahead to what's next, and out-of-the-box problem solving.

This separation of capabilities between the technological and the human is happening now. When a new technology takes on particular processes that people perform, we often end up with welcome changes that reduce the need for humans to handle repetitive, noncreative tasks— "taking the robot out of the human," as I like to describe it. This leaves people to focus on more important work— tasks that require deeper analysis, higher-level critical thinking, and an understanding of human interactions and relationships, including negotiations, transactions, and team building. When viewed in this way, technological developments can be embraced as a way to empower workers as innovators, thinkers, and advisors.

This is not merely a theoretical possibility. Evidence drawn from the early months of the post-COVID era suggests it is happening across a range of economic sectors.

Typically, recovery periods following recessions, epidemics, and natural disasters involve years of weak productivity growth. This is largely due to reduced capital investment, including spending on new technologies. But the opposite appeared to be happening in the first half of 2021. A research report published in March 2021—one

year after the acceleration of the COVID-19 pandemic in the U.S.—explored a number of trends around technology tools and services that are relieving many workers from tedious tasks and unnecessary commutes or travel. The report concluded that increased adoption of these tools could raise productivity growth in the U.S. and Western Europe by a full percentage point annually between 2021 and 2024. This represents a doubling of the average pre-pandemic growth rate and could generate growth in per-capita gross domestic product of about $3,500 in the U.S. and $1,500 in a European country like Spain.

Anecdotal evidence supports this projection. From restaurants to offices to factory floors, the pandemic inspired adoption of robots and artificial intelligence apps and programs that freed up workers from boring, often manual, tasks to focus on higher-value output. At the same time, cloud computing and videoconferencing tools reduced the often-unproductive time involved in office commutes and business travel.

In the hard-hit food processing and food services industries, AI was working overtime to automate specific tasks. A few illustrative examples from the COVID-19 period:

- A restaurant owner in California reported that he was using robots to transport food and dirty dishes between his kitchen and dining area, providing wait staff more time to interact with customers.
- Managers of a meat processing plant that was having trouble keeping staff during the pandemic turned to digital tools to replace workers, including

a newly developed camera system that uses AI to detect foreign objects in the meat.

- A fast-food chain in Ohio installed an automated voice ordering system, which the CEO claimed never failed to upsell customers on additional items from the menu. He also noted that the system never called in sick. The CEO declared that his chain would never go back to its traditional ordering system even after the pandemic was over.

Most recent college graduates and young millennials, having grown up as digital natives, seem to understand intuitively the benefits of new technologies in reducing workplace tedium and wasted time. In my work experience, I've found them to be excited by the prospect of working alongside disruptive technology. They're eager to offload repetitive tasks—which in some jobs can take up a large percentage of their time at work—and to spend more time and energy doing work they enjoy. They're also eager to develop new skills that will make them more relevant and valuable as employees for years to come.

However, using technology in ways that enhance rather than diminish human capabilities is sometimes easier said than done. Not every employer is enlightened about the skills needed or about the qualities to look for in their future management stars. Organizations of any kind that hope for long and productive futures must do their own homework to understand changes in their industry and what will be required of their employees if the organization is to survive and prevail.

The process of preparing for the future starts with our educational environments at all levels. In order to succeed in a rapidly shifting work environment, new graduates will need to be high-quality, high-potential, well-trained "professionals" as early as possible in their employment experience. To make this happen, employers must work with secondary schools, trade schools, and universities to help shape curricula and to build relationships with students earlier. Through this kind of collaboration, we can advance knowledge and skills for the future, and prepare students for what I call being "day-one ready."

As early as 2017, EY began a journey that would enable the firm to hire college graduates equipped with future-based competencies. We created a program called Day One Ready. With the University of Dayton as an early partner, we began working with colleges and universities from which we recruited talent to add future-oriented courses on topics like Lean Six Sigma and negotiation skills to their curricula. A program like this can enable new graduates to enter the job market better prepared for tomorrow's challenging careers, while participating colleges and universities can use these new offerings to build and differentiate their brands.

While young people generally embrace new technologies, experienced professionals and middle managers may view them as a threat. Some veteran professionals may think they're too set in their ways to easily pivot and take on an entirely new role in working with technology. But making this shift in attitude and knowledge is essential—and it can be done, especially if employers provide support.

In 2019, EY took its own upskilling efforts for current and rising executives to a new level. In collaboration with the Mendoza College of Business Executive Education at Notre Dame, we launched a first-of-its-kind program that included an intensive course called Disrupt by Design. It provided more than 20 of our professionals, all with five to 25 years of tenure, with training in topics like design thinking, agility, and Lean Six Sigma.

Organizations must find ways to ease workers' concerns by encouraging them to stay informed, then engaging them in decisions around the implementation of new technologies. Higher-level employees can use webcasts, audio programs, book summaries, online seminars, and conferences to stay on top of developments. Guidance and storytelling can help ease the transition. In some companies, "reverse mentoring" is being used, in which younger employees offer technology coaching to older workers.

The need to adapt to technological transformation is not brand-new. I remember that, when Microsoft Excel first came out, many employees with jobs that involved financial analysis worried that its functionality would render them irrelevant. Over time, they learned that the digital spreadsheet actually made their work much easier, enabling them to deliver deeper levels of analysis quicker and with greater accuracy. The human desire to satisfy our curiosity by expanding our frames of reference and increasing our problem-solving capabilities can go a long way to making us more comfortable with technological change.

This type of ongoing education, or "upskilling," can provide immense opportunities for experienced professionals to grow their impact often with less effort, not more.

Joining the Digital Natives

WHILE RECOGNIZING THAT THE ADVENT of smart technologies will make the skills that are unique to human beings even more valuable, let's also recognize that the on-rushing stream of technological innovations does create demands upon us. Because technological disruption will impact us all, employees and managers at every level of the organization have a role to play, and all must do their part to prepare both themselves and the broader business for change.

If you are in a management role, to help yourself and your organization succeed in the face of continued disruption, find ways to be an ambassador for change. Stay informed and excited about positive new developments. Encourage colleagues, as well as family members and friends, to learn about new technologies and the future good they might represent. Take some of the same courses your teams and mentees are enrolled in. Leaders can and should be champions of workplace evolution, setting the example as they guide and help their teams adapt to a new, digital workplace, fostering a commitment to lifelong learning. To borrow the famous line (often attributed to Mohandas K. Gandhi, though evidently never said by him), "Be the change you want to see."

This is a mission I've taken on personally. For me, it started with surveying the business landscape in 2016 and researching what would be demanded of people working in a business world increasingly influenced by the accelerated rate of new technology evolution and adoption. The facts I learned intrigued me. For example, a 2017 report predicted that, by 2020, the demand for data scientists and data engineers would grow by 39 percent. An overly rosy forecast? Not at all. By early 2020, the Dice Tech Job Report had found that the growth rate for data scientists and engineers had actually averaged about 50 percent year over year. We underestimated the impact all along! But there really had been no precedent to guide us.

In retrospect, this should not come as a surprise, as we've watched for several years the accelerating rate of new technology evolution and adoption and the downward trend in technology cost. As one result, the basic definition of how specializations like data science and analysis materialize into actual jobs is broadening. Jobs such as "systems analyst" and "computer systems engineer" continue to exist—but today you might also see a procurement manager, budget analyst, or marketing director using many of the same skills as well. Thus, the number of positions that fit under the "data science" rubric continues to increase faster than the overall number of job postings. It's a sign of the times: It's all about the data!

The moral of this story? The world needs more workers with a range of skillsets relevant to the demands of tomorrow's organizations. We need people who can build and repair technology systems, collect and interpret data, and design new applications. These skills are often described as drawn from the STEM fields—science, technology, engineering, and math. But the future isn't only about the technical work of extracting and storing the data. It's also about learning to access, interpret, analyze, visualize, and use all the data that the builders, coders, and extractors are capturing. Those tasks will fall to professionals as wide-ranging as economics professors, computer modelers, transaction negotiators, human resources professionals, corporate strategists, brand experts, project managers, and staff trainers.

What's emerging is a set of ecosystems with branches in various environments—business, communities, academia, government, healthcare—focused on translating technological capabilities into actionable solutions. Given the increased sophistication of human work and the desire of younger workers for meaningful careers, we are seeing more emphasis on purpose-driven work environments. The necessary trade-off is for individuals to do their part by adopting a mentality of continuous learning and a career-long commitment to upskilling, beginning in college and continuing up to and beyond retirement.

The work area I know best is the professions. When I think about the entry-level job skills that professionals will need to meet tomorrow's business needs, I'm amazed by the differences from when I graduated, or even from what was required less than five years ago. Consider this short list:

- A working knowledge of coding
- Data analytics and data visualization

- Agile, creative, design-oriented thinking
- Understanding of micro- and macroeconomics
- Basic understanding of AI and intelligent automation
- Awareness of evolving technologies like cloud computing and blockchain
- The fundamentals of business law, including emerging issues surrounding cybersecurity and data privacy
- Awareness of problems like the potential for unconscious bias in artificial intelligence technologies
- Sensitivity to issues of business ethics and their application to everyday decision-making

There will be continuing demand for people with specialized expertise in all of these tech-related fields, of course. But even those of us who are *not* tech specialists will need to have a basic understanding of these technological topics and how they impact our organizations. You may not make a living as a coder, an interface designer, an AI expert, or a cybersecurity specialist—but there will probably come a time when you'll need to work on a project with people like these. You'll want to know enough of their language to ask the right questions and to be able to explain what you need so they can help to provide it using their advanced technological prowess.

Technology Skills and Social Equity

IT'S ALSO IMPORTANT TO SURFACE the connections between technological skills and social equity.

Every March, when Women's History Month comes around, we are reminded of one area where career parity remains a steep, uphill climb: the continuing relative scarcity of women in STEM. Evidence is provided by these stats from Catalyst Research:

- Women account for only 29 percent of those employed in scientific research and development globally.
- Females are less likely to enter STEM careers than their male counterparts—and those who do launch such careers are more likely to leave them (53 percent of women compared to 31 percent of men). Many women who do leave STEM careers cite isolating or difficult work environments as the reason.

These data reflect an unmet need, an opportunity to be seized, and a challenge to be tackled. Leaders in many arenas of society, from business and academia to government, need to find ways to attract more females to educational and employment opportunities in science and technology fields and to ensure a conducive work environment for retaining them once they are engaged. This is important not only because STEM provides a fast track for opportunity and success today, but also because women offer a healthy diversity of perspectives and experiences that otherwise are missing from projects, programs, and leadership roles driven by STEM-related enterprises. Studies show that women tend to bring higher levels of empathy, engagement, and relatability to their work—all of which are important ingredients of critical thinking. These qualities are also keys to resolving some of the issues we currently experience with designing and managing social media platforms and artificial intelligence programs.

It's important to highlight the often-overlooked success stories of women in STEM or STEM-related fields, which could inspire more girls and women to consider similar career paths. Here are just three of the impressive women I've had the pleasure of meeting over the few years who have made a very real difference in science, business, and education:

- Dana Suskind is a practicing surgeon and professor of medicine at the University of Chicago as well as director of the Pediatric Cochlear Implant Program at the university. Her breakthrough work on implant techniques has enabled hundreds of children to gain or regain hearing. In addition to her medical practice, Suskind has also applied her learnings around hearing and language development in founding the Thirty Million Words Initiative, an evidence-based intervention program designed to minimize early language disparities among children from disadvantaged backgrounds.

- Gregg Renfrew is an entrepreneur who, after launching and selling a series of successful companies over the past 20 years, was inspired to apply her innovation and management skills in a more socially active way. In 2013, motivated by increasing evidence of links between beauty products and infertility, cancer, and other health problems, Renfrew created a skincare line that uses nontoxic ingredients, serving consumer needs while helping to protect the health pf people and the planet.

- Marsha Lovett is deeply involved in developing a deeper understanding of how students learn and then applying that knowledge to create improved educational technologies and teaching approaches. As director of the Eberly Center for Teaching Excellence and Educational Innovation at Carnegie Mellon University, Lovett applies theoretical and empirical principles of cognitive psychology to help improve teaching. She has published more than 50 highly regarded articles as well as a book on learning and "smart teaching" techniques.

There are many other women who are making a significant difference in STEM fields. They include Fei Fei Li, a computer science professor at Stanford, who is one of today's leading researchers and developers of artificial intelligence, working to imbue AI algorithms with empathy and other human sensitivities as apps and other software play an increasingly important role in decisions that directly impact human lives; Yewanda Akinola, a mechanical engineer who has developed water and sanitation systems for underserved countries in Africa, the Middle East, and East Asia, as well as founding the Global Emit Project to encourage young people to enter engineering—with a special focus on motivating girls; and Jennifer Doudna, a member of the departments of Molecular and Cell Biology and Chemistry at UC Berkeley, who was a leader in discovery of the revolutionary gene-editing tool, CRISPR, which allows scientists to alter DNA sequences and modify gene function, creating the potential to eradicate previously incurable diseases.

Female leaders like these represent a vast human resource that our society has not yet mobilized to its fullest potential. We need thousands more like them—and in the years to come, I think we will see them continue to emerge.

We need to focus also on increasing the level of racial and ethnic diversity among leaders in the STEM fields. This is not just a matter of equity; it also has major practical implications regarding the effectiveness, accuracy, and value of the technological tools we are developing and implementing.

Artificial intelligence, as just one example, has become an integral part of our daily lives, often unnoticed on the surface. But it impacts so many processes, functions, and systems that it is essential to ensure it is as bias-free as we can make it. From credit scores and lending decisions to resume scanning, search engine results, and online advertisements—every second of every day, AI is exerting influence that can foster fairness or create inequities. Machine programming is only as balanced and unbiased as the humans who produce it. Even algorithms and data unbiased when created can "learn" bias from humans interacting with them over time. Particularly at risk are women, people of color, and global ethnic groups who are underrepresented in STEM fields. The same systems of bias that keep women and people of color from seeking STEM careers are also influencing AI output. This needs to change.

Science, engineering, and technology skills—capabilities long listed among the critical skills for future success—are more in demand than ever. Let's keep the pressure on improving the opportunities for underrepresented groups in these key fields until parity issues like those that are periodically raised during Women's History Month and Black History Month become parts of our history, not our future.

Lifelong Learning and the "Soft Skills" That Matter Most

TRAINING IN THE STEM FIELDS is not the only way we need to prepare for tomorrow's challenges. If we're serious when we say that the skills unique to humans are even more important than the skills that machines can develop—and I am serious about it—then we need to acknowledge that so-called soft skills like empathy, judgment, mindfulness, and creativity are just as crucial as hard skills like math and science.

According to the staffing experts at Indeed.com, while hard skills reflect a particular, measurable ability to perform a task, soft skills involve how well you are able to work with others—to form teams, to collaborate, to build relationships, to adapt effectively to change, to communicate, and to solve problems. Soft skills are the underpinning for helping an employee use his or her hard skills to their fullest extent.

Because soft skills are never perfected, but rather are in a constant state of testing and growth, today's students need to approach their education and future working lives with a continuous learning mindset. And it is incumbent on employers of all types to help. One way is by providing in-house or online courses, perhaps through special arrangements with academic institutions. For example, there are an increasing number of massive open online courses (MOOCs), or "open loop universities," through which participants can experience learning in virtual classrooms as well as in real-world settings where practical participation and cultural immersion are possible.

As our focus on technology in the workplace continues to increase, future workers may find it even more challenging to stay "present." It takes a special kind of awareness to be immersed in digital technology while remaining grounded in human qualities like empathy and compassion. Employers can help by offering mindfulness training, self-care seminars, and counseling designed to assist people in "staying in the moment." Programs like these can help workers learn to truly listen to one another during meetings and to focus on specific tasks at a deep level of understanding. This ability to stay in the moment is essential to collaboration, and I believe it will be a defining quality in tomorrow's most effective leaders.

Far-sighted academics are moving quickly to respond to this trend. Jimmy Williams, executive director of the Engineering and Technology Innovation Management

(ETIM) Program at Carnegie Mellon University, views his job as cultivating next-generation innovation leaders. The ETIM curriculum is designed to enable engineers and scientists to develop business skills and thought frameworks they can combine with their technical knowledge to create and capture value from new technologies. "Innovation is about the ability to take good technology and wrap it in a great business model," Williams says. The ETIM program helps people do just that.

The current era of continual innovation and rapid transformation of business models calls for open-minded, agile leaders who can prepare people for constant change and help them find ways to keep succeeding and growing. That means being able to work with their people to navigate disruptions, seek diverse viewpoints on best solutions, and develop market- and people-sensitivities that enable teams to anticipate and prepare for change.

Tomorrow's workers must also be ready to transition from working with computers to preparing for the inevitable fusing of humans and technology, replete with implanted chips, smart tattoos, and computer-brain interfaces—all without losing the human element. This will require a unique skillset that isn't part of the typical business curriculum. Employers will have to choose whether to try to recruit the rare "unicorns" that possess these competencies or take transformative action to upskill their current and future workforces through their own investments in educational support.

Fortunately, many employers have recognized the vital importance of soft skills in today's economy. For example, LinkedIn reported in 2020 that, among the skills companies' worldwide report needing the most, the soft skills of creativity, persuasion, collaboration, adaptability, and emotional intelligence were ranked at or near the top.

The good news in all this is there is an accelerating need for uniquely human skills; the bad news is we don't have enough students pursuing them. It's a balancing act. The "trend pendulum" that in recent years swung skill and degree decisions toward STEM is building graduates in those areas, but that effort now needs to be balanced with a greater focus on other essential capabilities, such as effective listening, relationship- and team-building, project management, and problem-solving.

The COVID-19 pandemic has only increased the emphasis on soft skills. According to discussions mba.com has had with career management professionals and recruiters over 2020-21, the "new normal" requires greater empathetic leadership abilities. The Graduate Management Admission Council's annual Corporate Recruiters Survey asked more than 700 global recruiters which of 18 skills are the most important for business school. Interpersonal skills were identified as important by 81 percent of U.S. recruiters, 79 percent of European recruiters, and 79 percent of Asia-Pacific recruiters. Results like this suggest that, in an increasingly virtual world, it will be even more important to be able to connect with people and have the emotional intelligence to understand their needs.

The *Wall Street Journal* has further highlighted the same concern with an article exposing educational shortcomings in helping students "learn to think." The article's findings triggered reader reactions from interested organizations, senior executives, and team managers. Every year, students at 200 colleges across the U.S. take a standardized test called the College Learning Assessment Plus (CLA+). The test requires the use of spreadsheets, newspaper articles, and research papers to answer questions, take a position on a topic, and provide feedback on an argument. The 2013-2016 test results showed that, in many cases, there was little or no improvement in student critical thinking after four years of study at many of the institutions involved. Academic experts and employers interviewed by the *Journal* said the findings indicate a failure of our higher education system "to arm graduates with analytical reasoning and problem-solving skills needed to thrive in a fast-changing, increasingly global job market."

This leads to three important reasons why I believe critical thinking should be at the top of the list of non-technical criteria for future success:

- *In order to effectively reimagine the future, we need a rich mix of thinkers, doers, and fixers.* This means we need individuals with diverse backgrounds and studies, including liberal arts, biological and social sciences, technology, economics, and law, in order to fully realize all the potential opportunities and benefits technological transformation offers.
- *In an era when machines are freeing up smart, talented people to focus on strategic, purposeful work, the crucial question becomes, "What can I do with more time?"* In his book *Deep Work*, author Cal Newport flips the narrative on the impact of technology in a connected age, celebrating the

massive benefits of cultivating a "deep work" ethic rather than bemoaning the negatives of distraction. Having more time, Newport says, will enable high-performing professionals to assess, analyze, visualize, improve, and innovate at a higher level. He calls such "deep work" the "super-power" of the 21st century.

- *The value of a keen ability to ask better questions cannot be overstated.* In advocating for more young people to consider degrees in the liberal arts, a *Harvard Business Review* article noted, "If we want to prepare students to solve large-scale human problems, we must push them to widen, not narrow, their education and interests. Of course, we need technical experts, but we also need people who grasp the whys and hows of human behavior." The article goes on to explain why knowing the right questions to ask hinges on recognizing the problem you're trying to solve in the first place.

Of course, critical thinking itself is a difficult-to-define skill. One educational psychologist defines critical thinking this way: "Thinking about your thinking, while you're thinking, in order to improve your thinking." That does kind of sum it up, but it doesn't do much to clarify it. Perhaps critical thinking resembles certain other complex attributes, like confidence, personality, or style—you simply know it when you see it.

But however we define it, critical thinking is a valuable asset for the future of humanity. It's not about what you know or how you know it, but how well you listen, think, assess, and then relate what you've learned to solving a problem or engaging and collaborating with other people. For example, choosing the right career direction is a critical thinking challenge that involves reviewing hundreds of pieces of data drawn from research, personal observation, conversations with a wide array of individuals, thoughtful analysis of one's own goals and resources, and much more. The path forward is not always clear, and finding the right solution demands that individuals do their own exploration of a large number of opportunities and challenges, narrow those down to the best options, and eventually choose the one that simply feels most right. The ability to solve this sort of puzzle is complex, multi-faceted, and also absolutely essential.

In the pages above, I've mentioned some of the ways that organizations can help employees develop and enhance these soft skills, and I'll delve more deeply into organizational responsibilities in the next chapter. But in the final analysis, it's up to us as individuals to seek out and take advantage of the educational opportunities we need. Be proactive about making learning a regular part of your life routine—after all, it's your future that's on the line.

Underused Paths to Continued Learning

WHEN WE CONSIDER HOW working women and men are going to be prepared for the careers of the future, it's easy for our thinking to settle into a couple of familiar grooves. There's plenty to discuss regarding the weaknesses of our national primary and secondary education systems.

There's also a lively ongoing debate about how to make college and graduate school education accessible and affordable for more students, as well as the related topic of student debt as a massive burden for too many Americans. These are all important subjects for policy makers and citizens to ponder. But in the process, let's not overlook some other vital but often-neglected sources of education and training that we could apply with much greater effectiveness.

One of these is the system of apprenticeship, which is currently used much better in countries like Germany than in the United States. According to the Department of Labor, there are only about 600,000 apprenticeships in the United States—fewer than one percent of all jobs. But there is a case to be made that this medieval concept is due to experience a renaissance. As we reimagine the future of work, embrace disruptive forces, and integrate humans and artificial intelligence, apprenticeships have comeback potential as a time-tested learning model to help us prepare the workforce of the future.

In a world where workforce requirements have never been more dynamic, workers must be more flexible, collaborative, and thoughtful than ever before. Herein lies the value of the apprenticeship—the defining step through which learners become doers, working alongside a seasoned expert.

Traditionally, apprenticeships were a way for inexperienced newcomers to learn a new trade. A couple of generations ago, mastery of a skill or trade through an apprenticeship could provide income and occupational satisfaction for an entire career. Today, the concept is evolving from something you do prior to launching your career to a transformative experience that can help you continue to succeed throughout your career. Millennial and Gen Z workers will need to master numerous jobs throughout their careers, as the traditional "climb the ladder" path gives way to a less-predictable career shape involving multiple sideways hops and even the occasional backward step in preparation for a later forward leap. Apprenticeship learning can become a tool for workers to use in evolving their careers and mastering new skills at a fast pace.

I would challenge you—no matter where you are in your career—to consider engaging in some type of apprenticeship as a way to learn a new skill or redefine the way you approach your work. Start by spending 30 minutes a day in an informal apprenticeship, learning a new skill or exploring the performance of a new process or service. For those interested in a more significant commitment, there's a U.S. government website featuring apprenticeship opportunities at more than 150,000 businesses across the country. Having mastered one, move on to the next. If nothing else, an apprenticeship can be a great exercise in moving out of your comfort zone.

Get the most out of your apprenticeships by:

- *Learning best practices from those who have mastered their professions.* The same logic behind why athletes compete against champions applies. Surrounding yourself with masters of your trade will allow you to learn from the best examples and develop the highest competencies needed in your field. Everyone approaches job tasks in a unique way, and there is so much to learn from watching and training with someone who is revered for their work.

- *Developing habits and allowing them to be broken.* During the apprenticeship, you'll have the opportunity to develop positive habits, establishing for yourself what works and what doesn't. At the same time, recognize that there will come a time when, under the pressure of changing circumstances and fresh challenges, you'll need to break your old habits and develop new ones. Even a high level of expertise can be pushed to the next level when you allow yourself to be disrupted.

- *Being creative and innovative.* As you're learning, push yourself to think creatively, try new things, leverage the latest technology, push your limits, and take calculated risks. Breaking down traditional barriers will allow you to experiment and discover new ways of doing things. And remember that, even if you're not in a tech field, leveraging the latest technology can help you redefine your work and improve the processes you use to achieve your personal and professional goals.

The old adage "Practice makes perfect" takes on new meaning during an apprenticeship. Learning a complex new skill is about repeating many times the series of simple tasks that make up the larger skillset. This process leads to "tacit knowledge," which is a deep-seated understanding and "feel" for what you do that is hard to describe but easy to demonstrate. "Each time one skill becomes automatic," as author Robert Greene says in his book *Mastery*, "the mind is freed up to focus on the higher one." The deepening of tacit knowledge liberates the mind. You are "on automatic," a state that enables you to look for new challenges and invent new ways of using or improving the skill.

Thus, apprenticeship learning is about practice and discipline that ultimately allow you to put your own creative spin on what you are doing. Perhaps that is why the model has bred so many of the world's greatest masters—and why it may lead you one day to share your knowledge with an apprentice of your own.

Another powerful learning platform that millions of people have benefited from is military training.

Extreme Ownership is a book on leadership authored by Jocko Willink and Leif Babin, two U.S. Navy SEAL unit commanders. The authors led the most highly decorated special operations unit of the Iraqi War and survived many seemingly impossible military missions. They know something about extraordinary challenges, meticulous preparation, consequential decision-making, and how it feels to win—and lose—in competitions where the stakes could not be higher.

As I read the insights offered by these two exceptional leaders, it quickly became apparent that many of the same principles that drive success on the battlefield also apply in the business world. Their lessons from the battlefield underscore the value military veterans bring to leadership, education, and problem-solving skills. Many of the qualities these two warriors highlighted could be characterized as soft skills, including a number that align with a vision for the workforce of the future—for example, critical thinking, the elevation of teamwork over ego, and an emphasis on focus and high-quality execution.

A few of the lessons Willink and Babin highlighted stood out for me, including:

- *The goal of all leaders should be to work them-selves out of a job.* In this sense, *Extreme Owner-ship* is the ultimate succession planning tool! Train, mentor, coach, and formally develop up-and-com-ing leaders so they can continually assume ever-higher levels of responsibility.
- *Recognize that leadership is both an art and a sci-ence.* Be open to new ideas and welcome feedback. Don't always look for formulas or exact answers. Most problems have more than one solution—even more so for daily leadership challenges.
- *Leaders must be comfortable in times of chaos* and able to act decisively amid uncertainty.
- *Accept that not every decision will be a good one,* so handle mistakes with humility and dignity. Own up to errors, and demonstrate to your team that everyone makes mistakes. What's important is learning from them.

The relevance of leadership lessons from the military to life in the civilian sphere raises the long-standing debate in the United States over how best to recruit and staff the young women and men who serve in our military and protect our nation. Many experts say that the current volunteer-only military is the best-educated, most well-trained, and fittest force in our nation's history. But some policy experts say our country—and our society—would benefit from instituting some form of national service that would engage the efforts of young men and women from every social stratum, whether in the ranks of the military or in other kinds of work.

Without necessarily advocating for a national service program, I would point out the potential educational and social benefits that such a program would provide. If a majority of our young people spent a year or two working on worthwhile national projects while learning and exercising skills like leadership, decision-making, and team-building, both they and our country would benefit enormously.

If you are a young person seeking a constructive way to launch your work career that offers enormous opportunities for learning and growth as well as service to a higher cause, you may want to consider a stint in America's armed forces.

Vitality, Well-Being, and Life Satisfaction in an Era of Unparalleled Longevity

SOME EXPERTS IN HUMAN BIOLOGY believe that the first human who will live to the age of 200 is alive somewhere on Earth today. We can assume that some of the newly emerging longevity will be borne out of new, life-extending developments in healthcare, from better diagnoses to enhanced surgeries, medicines, and other treatments. Perhaps advances in genetically based medicine will also play a role. So will lifestyle improvements from better nutrition and enhanced exercise regimens to reduced pollution of the air we breathe and the water we drink.

Even if the goal of a 200th birthday proves to be elusive, it seems likely that, on average, the current generations will enjoy a longer lifespan than any other cohort in human history. This is another demographic shift that demands serious thought and attention from all of us. If we hope to make the most of all those extra years, we need to think seriously about *how* we live.

Whatever the formula for a longer life, I believe in the importance of certain intangibles that can make a significant contribution to longevity, even in the face of potentially unavoidable calamities, such as intractable diseases and accidents. Those include a positive outlook, living with a sense of purpose, practicing healthy habits (eating, physical fitness), maintaining mental acuity, and just doing things that make you happy. I group all these efforts under the heading of supporting well-being. And to that I would add the overlay of *vitality*—pursuing life, happiness, and good health not in a spirit of resignation, but with gusto. Embracing life with vitality makes a big difference in keeping people open to new learning, new experiences, and new challenges.

The pursuit of well-being and vitality should be part of your career planning process. For example, when vetting potential employers, consider the compensation and benefits package offered. Just a decade ago, a typical package was mainly about salary, vacation time, a variety of insurance options, and perhaps a pension or savings plan. But over time, as life and work became more complex and fast-paced—and as the competition for attracting and retaining the best talent increased—benefits packages and other human resources programs aligned with them changed in ways that would help employees optimize their healthfulness, address daily living challenges, and enhance their performance.

Today a strategic focus on promoting vitality can be a powerful way for one employer to differentiate itself from the competition. I would advise job seekers to ask about what an organization offers to contribute to their vitality and well-being. Look for evidence of an organizational commitment to self-actualization, to enabling people to bring their authentic selves to any situation. If such evidence is lacking, chances are that the organization's concern for the individual's "best self" is not there.

The challenges and opportunities being presented by our changing world are not all related to economics. Some connect directly with those elements of existence that make life worth living. As you plan your career in a society being transformed by technological, social, and demographic shifts, remember those human values and look for a personal path that will enable you to experience them to the fullest.

SPOTLIGHT

Coming Generational Shifts in Consumer Attitudes and Behaviors

Technological advancement and modern modes of consumption have grown up together. In the late 19th and 20th centuries, national and international markets, along with the giant consumer-goods companies that serve them, were made possible by new technologies of communication and transportation—cars and trucks, airplanes, radio, television, and even the ubiquitous standardized shipping container. More recently, the internet and the smart phone have profoundly changed where and how people shop and how payments are made, making the acts of buying and receiving goods more convenient than ever.

Now, 25 years after the launch of the first online super-retailer, another round of new technologies is transforming the process of consumption. In the years to come, artificial intelligence, augmented reality, and virtual reality are likely to play key roles in influencing marketing and order fulfillment. The implications for consumer goods manufacturers, wholesalers, and retailers will be profound, affecting everything from production processes and supply-chain costs to customer service and the nature of competition.

At the same time, it's important to recognize that consumer behavior is deeply interconnected with social, cultural, and even political realities. Modern consumerism has been associated with such trends as the rise of an affluent middle class, urbanization, near-universal literacy, expanded access to higher education, increased leisure time, and a growing appreciation for the values of individual freedom and self-expression.

Now a new set of trends is taking shape, driven by values that are characteristic of today's rising generations and potentially intensified by our post-pandemic environment. They include a decreased interest in acquiring products for their own sake; disdain for material excess and even the notion of ownership generally; a heightened focus on eliminating waste; and a growing interest in consumption practices that are socially and environmentally responsible. These attitudes will make marketing and selling to tomorrow's consumers more challenging than ever—which means that those companies that crack the code can expect to reap enormous rewards in terms of market share and profitability.

Let's delve a little more deeply into some of the big shifts in consumer behavior that we can expect over the next couple of decades.

- *The impact of smart technology will be a given.* Smart technology—in computers, tablets, cell phones, and a growing number of other devices—is increasingly becoming such an integral part of who we are that it feels invisible. In the years to come, we will find ourselves thinking about it less and less; instead, we will simply expect it to be there, performing for us and making life easier, faster, and more friction-free. Online shopping and ultra-rapid home delivery—which became necessities for millions of people during the worst of the COVID-19 pandemic—will be the new normal for countless consumers, who have gotten accustomed to being able to instantly compare the features and prices of competing products, to quickly scan dozens or hundreds of reviews from others shoppers, and to switch brands at the click of a button.

- *Consumer trust will be at a premium.* At the same time that customers are growing used to an ever-increasing level of quick, easy, seamless service, they are becoming intensely conscious of the importance of global supply chains and the user interfaces that link us to them. Over the last few years, we've all discovered how reliant we are on these technology-driven connections, especially as we've experienced unexpected shortages of vital goods during the COVID-19 pandemic and suffered the impact of cybersecurity breaches that have made our personal data vulnerable. Breakdowns like these are giving rise to increasing customer demands for reliability and a reason to trust. Companies that are best prepared to respond to those demands will have a big edge over their competitors.

- *Consumers will seek out goods and services that express their deepest values.* In the affluent nations of the world, the second half of the 20th century was largely about buying things—the cars, home appliances, electronic gadgets, clothes, and other lifestyle amenities that were suddenly affordable and available to millions of people for the first time. By the time the 21st century rolled around, many middle-class homes were overflowing with material things (hence the exploding demand for self-storage units in which families could stash their unneeded extra "stuff"). As a result, the drive to buy more things has slowed dramatically. Instead, many people are choosing to dedicate their discretionary income to services, experiences, and lifestyle choices that express and fulfill their personal values. Shopping decisions are increasingly influenced by social, cultural, and political priorities, with many customers seeking out suppliers they consider environmentally conscious, socially responsible, and culturally authentic. Brand images, logos, and companies that are perceived as being "on the wrong side of history" in regard to issues from racial justice to climate change will struggle to maintain their customer bases.

- *Consumer demands for transparency on the part of the companies they patronize will become even more intense.* As consumers seek to investigate the business practices of companies they buy from, those that offer consumers transparency about their processes and practices are likely to gain a meaningful market advantage. Food shoppers, for example, increasingly want to know everything about a product on the shelf, including where it comes from, the ingredients it contains, its nutritional value, the way its workers are treated, and the supply chain systems that deliver it to the store. Clear, accurate labels will be demanded—but so will access to information through many other channels, including social media platforms, efficient customer service departments, and company spokespeople ready to respond honestly to any new challenge or controversy.

- *Traditional brand loyalty will continue to weaken.* Again, the dramatic events of 2020-21 have helped to accelerate trends that had already been under way for some. Thus, pandemic-related inventory shortages caused many consumers to try alternative product lines; a 2020 study cited by *Forbes* magazine reported that more than 25 percent of respondents said they had begun switching brands more often than ever before. But the decline in brand loyalty had been under way for years, fueled by many factors, including the ease of brand-hopping via online shopping, the ubiquity of unbranded and store-label products, and the rise of subscription services for a wide range of goods and services, which allow consumers to meet their daily needs while making fewer long-term commitments. Companies in every sector will find themselves investing more to enhance the customer experience in an effort to shore up the brand loyalty that is becoming ever more difficult to maintain.

- *The shopping experience will continue to evolve.* With AI tools, recommendation engines, and shopping bots playing a growing role in curating the goods and services we buy and carrying out purchases, shopping is likely to become even more socially-shaped and experience-based. The 2020 *Forbes* survey cited earlier showed that more than 40 percent of respondents reported that they'd visited physical stores less frequently during the era of

COVID—and more than a quarter said this behavior shift is unlikely to change back even after the pandemic has passed. In response, companies that are deeply invested in physical locations will need to offer special benefits to reward in-store shopping, such as live events and experiences based around product lines.

As the changes we've examined suggest, the next couple of decades will be an exceptionally challenging time for consumer-facing companies. Such businesses can differentiate themselves for today's and tomorrow's consumers by following the lead of entrepreneurial startups. They must be willing to take risks, open to fresh ideas, and ready to adopt an iterative, test-and-learn approach to market research and product development rather than seeking to perfect a new product before launch. Among the questions that company leaders will need to grapple with are these:

- In a world where shopping is becoming increasingly frictionless and brand loyalty is growing steadily weaker, how can companies escape the threat of being "commoditized" and forced to compete primarily on the basis of price?
- What must companies do to juggle the intensifying and sometimes conflicting demands from consumers who want both ultra-fast service at rock-bottom prices and guarantees that the goods and services they consume are being produced in the most humane, worker-friendly, and environmentally sustainable ways possible?

- How can traditional product-centric companies in industries like food, apparel, and household goods remain relevant to consumers whose interests are increasingly shifting their spending to experiences and services?

Consumer goods businesses will need to take on new and unfamiliar tasks, attitudes, and business models if they hope to survive, much less thrive, in today's ever more customer-centric world.

CHAPTER FIVE

The Challenges of Leading an Organization in a Rapidly Changing World

A S WE'VE SEEN, in the last 20 years, the pace of change and increasing maturity of data science, artificial intelligence, machine learning tools, cloud-based platforms, and more have moved the role of technology in the enterprise well beyond departments such as design, manufacturing, logistics, and back-office operations into white-collar offices and professional services. More recently, the COVID-19 pandemic has motivated big shifts toward remote work, forcing many organizations to rethink workforce models, numbers of employees, uses of workspace, and greater incorporation of technology in just about every imaginable way. To make the most of the new technological capabilities and the accompanying social changes, organizations must have both the right tools and the right people in place.

As organizations look ahead to their post-pandemic futures, many see a hybrid workplace model, combining some portion of virtual work with time in an office. The virtual model worked better than many expected in 2020, with numerous companies reporting significant productivity gains. According to the Associated Press, U.S. productivity for all of 2020 rose 2.6 percent, an improvement over a 1.7 percent rise in 2019 and a 1.4 percent uptick in 2018.

At the same time, however, lack of a clear vision of the working world going forward has prompted high levels of anxiety in workforces, with many employees reporting feelings of distrust and isolation that need to be addressed if the hybrid work model is to work well. Fast-paced adoption of digital technologies due to the COVID crisis could contribute to continued productivity growth, but other factors could depress gains, such as lower labor force mobility and lack of skills that align with needs.

What do leaders need to know to get out in front of the many challenges of this new age? I believe the best way to start is to get acquainted with both the technological changes and the evolving human capabilities that will shape our future. Even as smart machines get better at task performance, we will need intelligent, thoughtful, well trained, and highly motivated people to draw on their domain knowledge, to innovate, to make sound ethical decisions, and to ask the right questions at this pivotal time for business, society, and humanity. Transparency and effective communications can help organizations build resilience, loyalty, and trust.

Aligning People, Machines, and Transformative Processes

IN STUDYING A WIDE RANGE of companies that are grappling with the new wave of technological innovation, futurist Jacob Morgan identified the 12 habits of successful, highly collaborative businesses. The list began with understanding the need to focus on individual value before corporate value as well as the importance of putting strategy decisions before technology decisions. And strategy—today as always—is fundamentally about creating tools, conditions, and competitive circumstances that enable your people to be more successful.

Improved productivity and timely new services are important, of course, but most organizations today are, at their essence, people businesses, in which employees, customers, community members, and other stakeholders all play crucial roles in determining success or failure. Therefore, our objective in these changing times must be to find the best technology tools while simultaneously adding value for people. For me, that means implementing a talent-first strategy to help guide our adoption of AI and intelligent automation. This includes aligning business strategy and talent management, thereby ensuring that talent and operations speed along parallel and mutually beneficial tracks. After all, the best collaborative technologies and platforms in the world are useless if your people can't see the value these tools deliver and use those tools to pass that value on to their internal and external stakeholders.

Today's leaders can and should start now to look at the new wave of technological innovation with fresh perspectives on what it means for their people, including how collaboration with new technology tools can add value. With big changes arriving daily, companies that devise strategies that embrace innovation rather than resist it, and that inspire optimism rather than fear, are the most likely to claim the first-mover advantage and transform their industries. Position your employees to lead your technological journey, not fall behind.

In this chapter, we'll look at five keys to an effective response to today's new challenges:

- Making organizations more adaptable
- Developing leadership methods that connect with the rising generations
- Promoting continuous learning
- Understanding and addressing the challenges of remote work
- Building diverse, inclusive, and equitable organizations

Making Organizations More Adaptable

ANYONE LEADING TRANSFORMATION in an organization knows it is natural for people to resist and struggle with change. Thus, the first step in implementing a new technology vision must be to ensure that employees are buying in. Employees perform better through change when they understand why change is happening. In addition, today's younger generations—Millennials and members of Gen Z particularly—want a sense of purpose and mission around which to shape their careers, and they want to have influence over how they adapt. They grew up with sophisticated technology playing a major role in their lives, so they are intricately linked to technology as part of their identity.

The impact of today's disruptive forces—technological, economic, and demographic—is propelling a new look at professional skillsets, workforce models, and even how we deal with distraction, stress, and complexity. As a result, the ground rules for organizational transformation have evolved to include two new principles.

First, workforces are now developing a *collaborative* relationship with new technologies, a reality that affects how we choose technology tools and applications, how we introduce them to the workforce, and how we use them. The assembly-line robots and personal computers that distinguished the information age—when the internet was fully emerging at the turn of the 21st century—were not the collaborative machine partners of today. In decades past, people worked *on* machines, programming and keyboarding them. Today, intelligent automation and artificial intelligence are about working *with* machines. We are now actually partnering with these tools and seamlessly integrating them into the flow of our work.

This has turned out to be an important distinction. Collaborative technologies are effective only to the extent that people accept and use them. Thus, as we stand at the brink of increasingly fusing people and technology, organizational processes and cultures will need to reflect a more flexible and adaptive approach to both people and technology.

Second, innovation has now become an essential workplace activity for *everyone* in an organization, not just for a handful of technology or strategy specialists.

"What can we do to inspire more innovation?" This must be among the most commonly asked questions in business today. Find the answer, and you've got the secret sauce that thousands of technologists, engineers, entrepreneurs, marketers, and management executives spend millions of hours every day trying to discover. After all, in today's fast-changing world, it's no longer good enough to figure out what's next. Today, the quest is to identify what's next *after* what's next.

With innovation as the most elusive riddle of the 21st century, imagine how surprised I was to get great advice on the subject from a most unlikely source—one of the world's leading comedians, Jerry Seinfeld.

Seinfeld was asked by *Harvard Business Review* how he came up with the concept for his talk show, *Comedians in Cars Getting Coffee.* Thinking up something new and valuable is easy, Seinfeld responded: Just ask yourself the question, "What am I really sick of?" Seinfeld explained that he was tired of watching formulaic talk shows and hearing people exchange self-serving compliments while promoting their latest movie, play, or book.

I was impressed by the simplicity of Seinfeld's approach to innovation. And while I tend to take a positive, optimistic view of life and business, there is something in the *negativity* of Seinfeld's innovation theory that makes it particularly compelling—and effective. It builds the process of innovation around the avoidance of something unpleasant and offers us the emotional involvement we experience when we have the opportunity to turn something bad into something better. This is what really makes the Seinfeld philosophy of innovation work.

Isn't it true that some of our most daring decisions are born out of a need to change our circumstances—or out of a need to get away from something that is really dispiriting or distasteful? Consider the reason behind the smartest job move you ever made, or the best idea you ever contributed to a brainstorming session. There's a good chance it had roots in something you were sick and tired of. Frustration can propel a fresh way of looking at a problem.

Negative experience has spurred innovation in fields way beyond Seinfeld's world of comedy. Leading technology innovators have found inspiration in problems from the world around them. Michael Dell has said he was inspired by asking why a computer should cost five times as much as the total cost of its parts. His resolution of the problem resulted in a revolutionary new business model. Scott Cook had his idea for personal financial software after watching his wife tire of trying to keep track of their family's financial paperwork. Meg Whitman has said of the many entrepreneurs she has worked with, "They get a kick out of screwing up the status quo. They can't bear it. So they spend a tremendous amount of time thinking about how to change the world."

Notice that, to turn an unhappy experience into a driver of innovation, we must learn to embrace the discomfort that is naturally involved in dwelling on things we dislike.

Of course, embracing discomfort doesn't come naturally to us. Comfort is key to some of life's basic needs: a good night's sleep, an enjoyable meal. But sometimes, too much comfort is not in our best interest. Too much comfort—and the complacency that usually accompanies it—is the enemy of fresh thinking and constructive change. As hard as it may be to step away from comfortable circumstances, discomfort drives new ideas—which means it's essential to surviving and thriving in this decade and beyond.

How your organization adapts to demographic shifts, increased globalization, and technology transformation truly matters. That requires you to embrace the discomfort you may feel when you force yourself to think hard about the difficult disruptions these changes often produce.

Notice, too, that some of the traditional assumptions surrounding business disruption also need to be challenged. While conventional wisdom holds that true disruption is generally driven by newcomers and upstart challengers, long-time incumbents also can be accomplished disruptors. How? By staying alert to signs of change and opportunity and then quickly capitalizing on what they've observed by being open to new ways of thinking and doing things.

The leaders of today's most innovative companies practice a few basic principles that enable them to remain highly adaptive even as their organizations grow bigger and more complex:

- *They continuously seek out new ways of thinking.* The rule to follow: Never get complacent, no matter how far out in front of the competition you may be. Keep an open mind, acknowledge new facts and realities, and stay out in front of new trends and market dynamics. Listen, read, participate!

- *They study what others are doing and seek to learn from it.* Consider convergent business strategies that could make your organization more agile and adaptive. These could include expanding into adjacent businesses, making acquisitions to obtain access to talent, and merging with a competitor or with a small, but proven, disruptor. Look for opportunities to form partnerships with other organizations that will enable both sides to gain complementary knowledge or capabilities.

- *They foster a culture of innovation in their organizations.* Introduce policies and practices that promote fresh thinking. Hold competitions, offer incentives for great new ideas, bring in fresh talent from outside to "shake up" thinking, and reward behavior that monitors disruptive trends. Encourage employees who care about social issues or who are known risk-takers to speak up, helping them to reshape parts of your organizational culture that do not respond well to change.

- *They constantly update their thinking about strengths and key competencies.* Virtually every industry has experienced so many changes in recent years that traditional norms and expectations regarding skills, personal qualities, and other hiring criteria probably need a thorough overhaul. Reexamine your handbooks, recruiting ads, orientation materials, and other documents, and update them to attract the people you need today and for the future.

- *They seek out diversity in all its forms.* Diversity and inclusiveness are no longer only about gender, ethnicity, and other traditional markers. They're also about how your organization thinks, engages people, and leverages new and different ideas.
- *They aspire to become tomorrow's visionary leaders.* One of the most important qualities of your executive corps is the ability to see around corners—to anticipate what's next for your company and its industry. Strive to foster forward-looking behavior in your organization's top leaders, and engender an appreciation for the unexpected and unexplored.

In short, to win at adaptability, get comfortable with being uncomfortable! Encourage your people to be bold disruptors. Embracing change will strengthen your ability to make problems and challenges into springboards for positive new developments. In recent years, for example, the firm I work for opted to upend the traditional approach to internships by inviting interns to disrupt our thinking rather than support it. We get great new ideas, and the interns love this new type of challenge

How can you ensure your organization is dealing effectively with current challenges and the dynamics of transformation—what I call "future-proofing"? Here's an exercise you may find useful as a starting point.

Sketch a table with the three driving forces discussed in chapter two—globalization, pace of technological innovation, and demographic shifts—listed vertically. Depending on your industry and your own perspective, you may choose to add a couple of additional driving forces that you think will be particularly influential in the years to come. Then jot notes beside each force about its likely impact on your market, your industry, and your own organization.

Then set about future-proofing your organization by thinking about what you need to *stop doing*, to *start doing*, or to *do differently or better* in response to each impact. Like fire-proofing, earthquake-proofing, hack-proofing, or any other protective risk-management activity, future-proofing should be a routine practice for organizational leaders.

Admittedly, it's not always easy to predict the impacts that future trends will have. The track record of economic and investment forecasters is rather poor, especially when it comes to predicting crises. In his book *The Black Swan*, statistician and scholar Nassim Taleb found a metaphor in nature—the infrequent occurrence of black swans—that he uses to designate rare, seemingly random events with enormous consequences. The fall of the Soviet Union, the terror attacks of September 11, the rise of the internet, and the financial crisis of 2008 are examples of black swan events—paradigm-shifting occurrences that few people foresaw.

But imagining possible black swan events of the future—and preparing for them—is not impossible. Most often, the data needed to foresee these upheavals is available, obscured by other factors: poor management oversight, communication failures, a narrow focus on day-to-day problems, or sheer lack of imagination.

A commitment to agility, open-mindedness, and diverse thinking can help you do a better job of imagining what tomorrow may bring. We are all living in the same roiling ocean of change, but only a smart, alert few will continue to lead their professions and markets. Taking practical steps to enhance your organization's adaptability quotient should be high on your agenda.

At the same time, it's not always necessary to be the first to recognize an emerging trend. History shows that, in many cases, first movers in specific industries have not maintained their number-one positions. These "pioneers," as author Adam Grant calls them in his book *Originals*, have often lost ground to later-arriving organizations he calls "settlers." Few people remember that the first video game console—the Magnavox Odyssey—was brought to market way back in 1966, only to be competitively overtaken 10 years later by Nintendo, which used that time to invent more user-friendly gaming options. Something similar happened when pioneer Blockbuster was swamped by settler Netflix, which used the internet to build a better system for home movie rental. Grant concludes, "Being original doesn't require being first. It just means being different and better."

There are exceptions to this pattern, of course. But there is something to be said for letting an innovation get absorbed into the culture, observing market shifts and changing generational attitudes, and adapting accordingly. It's never too late to recognize and respond creatively to an ever-changing world.

Developing Leadership Methods That Connect with the Rising Generation

IN 2020, MILLENNIALS, born between 1982 and 2000, were about 60 percent of the workforce. This digital-native generation has a head start when it comes to coping with the changes that technology and other driving forces are producing. They were born into the digital world and intrinsically possess a digital mindset, two advantages that help them address the challenges of keeping up with ever-evolving technologies. But for businesses, the benefits of this innate sensibility in today's workforce go well beyond the savvy use of technology and social media. Digital natives bring a competitive advantage to organizations by challenging traditional thinking at a time of great disruption, and their natural self-confidence and digital mindset enables them to quickly grasp and adjust to new developments in the business landscape.

However, as many companies know, adapting an organization to better attract a digital-centric workforce is not easy. For many large businesses, the task can seem insurmountable. Fortunately, there are a number of levers management can begin pulling today to respond to the needs and demands of the growing digital-native workforce.

One lever has to do with the rules that govern workforce management and career development. A 2019 FlexJobs survey with 7,300 respondents revealed that flexibility issues now play a big role in employees' job choices. This factor is driving changes in human resource benefits and practices that include more flexible work schedules

and leave policies. Among other new benefits being offered by leading employers are reimbursement of com-
muter costs; confidential counseling services; financial
planning advice; wellness programs; professional development courses; pet insurance; and sabbaticals, personal
leaves of absence, and parental leave for employees of all
genders. Where possible, consider incorporating these
kinds of new benefits into your company's menu of offerings to show that you are listening to the needs of today's
employees and are serious about meeting them.

As the workplace evolves, so do the values and goals
of workers themselves. *New Republic* magazine reports
that "climbing the ladder" is not as typical a path today as
it once was; younger workers have become more independent and entrepreneurial in their thinking about how
they map their careers. The same FlexJobs survey refer-
enced above found that 30 percent of working professionals have left a job because it didn't offer flexible work
options. Sizeable fractions of survey respondents further
noted that work-life balance (75 percent) and meaningful
work (55 percent) are significantly more important to them
in considering a job offer than more traditional benefits
such as health insurance (39 percent) and 401(k) contributions (32 percent).

New technologies are also giving organizations the opportunity to rethink how they deploy people as a workforce. Alongside the traditional, nine-to-five, in-the-office
workforce, we can now add "anywhere, anytime" working
arrangements, accommodating virtual and contingent employees, contractors, gig workers, and crowdsourcing con-
tributors. These new ways of working make a great fit with

Millennial and Gen Z preferences for more flexibility in their work lives and the opportunity to explore new career paths for growth and success.

On the other hand, new technologies have also brought some challenging sociological and psychological issues, including a major increase in distractions and stress. There's a lot competing for our attention, and some research suggests that the average human attention span has declined by 30 percent since 2000—down to eight seconds of consistent focus from the previous 12.

As we face new problems like overstimulation and hyper-connectivity, business leaders are increasing the importance placed on clarity, deep thinking, and reflection. Creative work requires focus, intense periods of cognitive effort, and an unfettered ability to "go deep," which many mind-clarity experts say is essential for mastering complicated information. Many companies are already making policy changes to support such thinking, ranging from the introduction of wellness and mindfulness programs to benefits programs that include personal and family support services.

Transparency and open communication are also practices that today's rising generations value. To meet this demand, many companies are implementing new forms of information exchange and employee coaching. For example, leading retail and healthcare companies have implemented reverse mentoring programs that open up new lines of communication between younger and older generations of employees.

The rising generations of employees want to make a real difference at work. Unfortunately, some consider their jobs to be devoid of value and meaning. When they do, it has a deleterious effect on their levels of loyalty and engagement. The impact on organizations can be painful; for example, it's estimated that Millennial turnover costs employers in the U.S. $30.5 billion annually.

The flip side of this trend is that companies have an opportunity to attract and retain digital natives by offering them the opportunity to pursue challenging and stimulating work with a positive impact on the world. Keep digital natives motivated by activating a sense of purpose throughout your organization, from corporate strategy and branding to operations and talent development.

Digital natives also have a tremendous thirst for knowledge. Practices like continuous learning, upskilling, feedback, and development can ensure they continue to expand their skills. Employees at Intel and Tesla, for example, use online learning platforms like Degreed, EdX, and Khan Academy to learn new skills, stay up to date on the newest developments, and sharpen their abilities. Instituting this kind of 360-degree learning program sends the message that you're invested in your employees and that your business knows what is important to them.

The skills, insights, and values of today's rising generations of workers can help empower your organization to take full advantage of the creative opportunities presented by new technologies. Yet most companies have so far failed to make this happen. For example, research reported in the 2018 Digital Transformation Barometer released by ISACA, a leading organization for business technology professionals, shows that less than a third of enterprises are making it a priority to consistently evaluate the opportunities that emerging digital technologies might provide.

Recruiting a cadre of digital natives and providing them with the kinds of work opportunities and challenges they relish can help your organization move ahead of the competition on this front. Take the first step with some simple investments, and you're well on your way.

Finally, remember that the changing shape of today's workforce also means that companies must accept the reality that the careers of our employees are not under our control. Sometimes empowering people means having to let them go when they feel called to pursue more fulfilling job opportunities outside our walls.

In 2021, I did a webcast for Women's History Month with five women professionals who'd each made the decision to leave my firm over the past two years. I learned that, while their reasons for making a big career change were different, all five shared a trust in the firm's commitment to supporting them in their new ventures. It's rewarding to see people move up your organization's ladder, but it's also rewarding to watch them finding their own ways, wherever they may lead.

Today's world is increasingly complex, but it remains an opportunity-rich environment for those who put in the time and the effort to succeed. It helps to start out with an organization that puts its people first.

Promoting Continuous Learning

COMPANIES MUST HELP their current employees grow and upskill to learn new competencies and take on new roles. Companies may also need to partner with universities to help ensure that graduates are better prepared for the careers of the future. If employee development is part of your leadership portfolio, you should be asking yourself: Are our tools and training programs intelligently targeted for today's workforce? Do they provide the right forms of development based on current knowledge gaps, employee needs, and a variety of learning styles?

To foster the higher-level thinking skills—interpretation, intuition, problem-solving, analysis, innovation—that clearly differentiate humans from machines, organizations must:

- *Invest in domain knowledge.* In order to offer uniquely human value, employees must have a deep understanding of the organization, its products, services, and processes, as well as the environment in which it operates. This kind of knowledge can't be downloaded; it's gained only through experience and history. This human-based knowledge can be combined with tech-driven tools and algorithms to help generate meaningful change.
- *Keep up to date about what's happening in your industry and markets and in those of your customers and competitors.* Sometimes we may be inspired by changes we see in other industries. The key is

to keep learning—do a lot of reading, attend industry and technology conferences, and encourage ongoing training to deepen your organization's existing areas of expertise or to expand your scope into new territories.

- *Read between the lines.* Technological tools look at data. They spew out numbers, patterns, and calculations, but they are still mastering the art of understanding the reasoning behind the numbers. Humans can enrich the results by learning to combine tech-developed data and personal experiences or observations into actionable insights and strategy.

- *Leverage human connections.* At least for now, technology tools don't have feelings and can't experience empathy. Foster the innate human ability to yet connect with others in a meaningful way by forging relationships with clients and peers. Team up with people you regard as cutting-edge thinkers and build out your network to get broad exposure to new ideas.

- *Bring digital proficiency to every job.* Productivity skyrockets when our professionals integrate their domain knowledge and people skills with machine-proficient processes in an immediate and direct way. The future requires seamless connections between our knowledge of the client's needs and our ability to innovatively leverage technology for the best solution.

- *Help employees hone the soft skills needed to complement their technical knowledge.* Soft skills were the most sought-after capabilities cited by 65 percent of 650 employers in the 2019 Morning Consult study. (Quantitative and technical skills lagged behind at 47 and 50 percent respectively.) A similar study by LinkedIn noted that soft-skill demand remains largely unmet. Since many of the courses needed to foster soft skills are not yet offered by academic institutions, employers will be required to drive the way forward through hybrid learning programs with institutions, massive open online courses (MOOCs), and internal training.

- *Respect data privacy in an age of online learning.* Most companies strive to keep their data analytics and algorithmic capabilities proprietary even as the number of cyberattacks and data leaks grow each year. When a large number of users shift to an online platform, as has happened with the steep rise in virtual learning and teaming, the need for infrastructure that protects personal data increases. There are federal and state laws in place to protect personal privacy, but many states and academic institutions lack the funding, expertise, or personnel to deal with data privacy issues effectively. Businesses, educators, parents, and students must become well versed in how new technologies are being used, what protections are in place, and what needs to be done in the future to thwart cyber-criminal practices.

A workforce with the right mix of digital skills, domain knowledge, professional service competencies, and personal attributes—all complemented by access to the latest tools—can mold a future that produces enormous benefits for our customers and for society.

Recent college graduates may not realize that education doesn't end when they're handed their diplomas. Primary, secondary, and tertiary schooling are only a small part of the education equation for tomorrow's workforce. Employers play a critical role in continuing learning beyond formal education. They keep the ball rolling.

A few years ago, in my talent leadership role, a personal commitment to lifelong learning prompted me to create a special intensive learning program we called Elite Skills Week. It was a distinctive learning experience that brought participants together to develop, collaborate, and apply their new knowledge. We identified five future skills our professionals needed to grow; then we developed a learning experience around each one:

- *Mindfulness.* In today's world of constant distractions, work demands, and technology notifications, it's difficult to stay focused on complex work. Learning how to "be present" helps professionals maintain steady interpersonal interactions, concentrate on challenging, strategic tasks, and reduce time spent toggling between competing demands.
- *Lean Six Sigma.* This team-oriented methodology for process and operations improvement has proven very successful over several decades, though it has not yet been applied in every industry

or profession. It actually is a combination of two sets of principles, tools, and practices: Lean, driven mainly by waste elimination, and Six Sigma, which focuses on quality improvement. Together, they help teams get to the root cause of a business problem and follow a step-by-step process to resolve it.

- *Robotic process automation.* Professionals need to understand how to identify process areas that can be automated, allowing humans to focus on higher-value tasks.

- *Excel modeling.* As the data we use increase in volume and complexity, we need to master the tools and knowledge needed to organize, analyze, and report the data in a repeatable, time-efficient way.

- *Data visualization.* Once analyzed, data must be presented visually to convey complicated perspectives in a meaningful way. Data visualization helps workers see patterns and hone in on pertinent information that generates actionable insights.

The Elite Skills Week program included a combination of instructor-led courses, outside speakers, teaming activities, and plenary sessions. It's just one example of how lifelong learning can be implemented to help ensure that employees are equipped with future-focused skills.

Keeping up with new learning technologies can be a formidable challenge for businesses and educators, but it is a critical one for them to meet. Studies show that students want to use more technology in their training and that they learn better when they can access their lessons electronically. Technology-based learning and virtual reality can supplement traditional ways of learning, enhancing ongoing education initiatives to ensure the best outcome for all professionals. So, too, can *gamification*, which applies game-playing rules and competition to a learning activity, thereby encouraging interest and engagement, especially among generations that grew up with video games. Examples of gamification techniques include incentivizing student progress by awarding badges for stellar completion of activities and creating challenges or quests that make learning more fun and memorable.

When it comes to upskilling, everyone in an organization should be included. Both new and experienced employees can benefit from training programs that will expand their mental horizons. Members of senior management may also need to get introduced to disruptive thinking concepts that will help build their awareness of the changes going on around us.

Ongoing professional development not only helps individuals stand out but also improves talent retention and company performance over time. This technology-driven era promises to deliver both new challenges and new opportunities for human resources leaders. If you start preparing your future workforce today, your organization can be ready to seize the upsides of disruption tomorrow.

Understanding and Addressing the Challenges of Remote Work

DIGITIZATION IS IMPACTING when, where, and how employees work, creating change at an accelerating pace.

Expect the number of "traditional" workers to decline significantly over the coming years as we expand the other portions of our workforce, some of which are digital (robotic automation), some offshore, and some gig-based. For a huge swath of the American workforce, the idea of a "job" is giving way to independent employment as a permanent freelancer or subcontractor. In fact, a Gallup poll released in May 2020 showed that 44 million U.S. workers—28.2 percent of the workforce—were self-employed at some point during the previous year. It is likely that number increased during the pandemic of 2020-21 as traditional workers lost full-time jobs and many started their own businesses.

One also can argue that during the worst period of the pandemic—March 2020 to March 2021—digitization proved to be the salvation of many workers and companies. Businesses were able to rely on automation-enabled tools, such as artificial intelligence tools and programs, to help keep up a high level of productivity with fewer employees and largely virtual workplaces. New technologies emerged to facilitate work from home and other remote locations, including platforms for virtual meetings, streaming video production tools, cloud-based data exchanges, tools for document sharing, electronic payments, and digital signatures, among dozens of others. And as the pandemic wound down and employment needs began to rise, electronic job-matching sites provided the digital platforms to capture, display, and sort employee credentials and connect candidates with employers.

Another important recent trend is the rise of *digital nomads*—employees who travel the world while working virtually. Many digital nomads work as freelancers or entrepreneurs, some as programmers or bloggers. Some may work on a set schedule, others when it suits their traveling lifestyle. With the ability to connect from nearly anywhere in the world, these workers can stay tuned in with coworkers or managers as needed while maintaining their flexibility.

But as various forms of telecommuting and remote work continue to evolve, organizations must share responsibility for making work from home productive and positive for employees. Tips for effective work-from-home strategies from the Nonprofit Leadership Center include creating a dedicated workspace and establishing a routine that resembles a day in the office but also addresses personal and family needs, including scheduling breaks and times for physical movement. To be effective, these tips need to be supported by organizational policy and leadership commitment to both the challenges and the opportunities remote work presents. Consider creating a gift card or corporate discount program that helps employees buy office equipment or furnishings; give reimbursements on wellness classes and fitness equipment; and offer flexible work-hour options to the extent possible.

And remember the positive impact of simply expressing gratitude for employee resilience, commitment, and extra effort—verbally or in a note. "Thank you," when delivered with sincerity, are two simple, but very powerful, words.

When planning policies for remote work, don't forget about security. Employees have an obligation to protect

company data; the company has a complementary obligation to protect employee data. As more information moves
to cloud-based systems, it's important to look at methods
for protecting employee privacy. As a first step, leaders
should ensure that any third-party vendors or partners are
taking the appropriate steps to protect their data.

As business leaders, it's our role to empower employees—whether they are local or a world away—to work
better together. The best technology solutions can help
keep employees connected 24/7. But it's even more important to solidify relationships among employees by crys-
talizing our organizational mission around a higher
purpose. Whether it's modest or far-reaching, such a purpose can have the power to unite the entire workforce as
well as providing a competitive differentiator.

Building Diverse, Inclusive, and Equitable Organizations

EMPLOYERS NEED TO BE READY for a new set of questions on diversity and inclusion (D&I) practices. Given the
overwhelming economic and social impact of the COVID-
19 pandemic, it's easy to overlook the other dramatic news
trend of 2020—the resurgence of concern about racial equity and America's troubling legacy of prejudice and injustice. For all its shortcomings, 2020 hopefully will be
remembered as a year in which mutual respect and empathy were energized through increased momentum for D&I
initiatives. This is especially important for leaders in a talent management function.

I saw this trend being played out during a visit to Norfolk State University (NSU), a historically Black college in Virginia, where I have the honor to serve as a member of the School of Business Board of Advisors. I'm sharing the following story because it's indicative of a shift in how the traditional employment interview is performed—and it illustrates vividly why employers should be prepared for the change.

It all started when Glenn Carrington, dean of the NSU School of Business, wanted to better understand how his students had been processing social unrest, and how they might, in turn, enhance their understanding of where potential employers stand on key issues. So he took a nontraditional approach to help business students who were interviewing for jobs assess which organizations "walk the talk" around inclusiveness and equity. Dean Carrington asked the students, "What are the questions *you* would most like to pose to the people you will be working for and with? Listen to the answers. Then gauge their credibility for yourself."

There was no hesitation from the students. Below is a sampling of questions from the discussion, which Dean Carrington also shared with the NSU Board of Advisors, comprising business leaders from a wide range of industries. The same questions can also be applied beyond racial minorities to other potentially excluded groups.

- "You told me about your diversity and inclusion policies, but now I'd like to hear about other African Americans who work here or have worked here. What was their journey like? How many are

in positions of senior responsibility? How many are on your board? What is your minority retention rate overall? Can I speak with a current or former minority employee?"

- "Do you hold your employees accountable for how they respond to social injustice? For example, are there penalties of some kind imposed on those who express racist or other discriminatory behavior? If not, how do you recognize genuine and positive diversity and inclusion contributions?"

- "How do minority employees develop relationships with non-minorities in your organization?"

- "Do both minorities and non-minorities act as mentors and advocates for strong performers, coaching or sponsoring them for new job assignments, promotions, and salary increases?"

- "I've always been told I should never discuss subjects such as politics and religion in the workplace. But I want to know before I take a job somewhere if you are committed to diversity and inclusion. How do you feel about discussing these topics with me?"

- "Will I be allowed to bring my true self to work—how I dress, wear my hair, celebrate traditions? Or is there a gold standard for really fitting in?"

- "If I am having what I perceive to be a 'racial' or other 'social justice' issue with my boss, will I be viewed as a troublemaker? Is there an independent person with whom I can speak confidentially about the matter? How does that work?"

- "Can my journey and current circumstances help the company do a better job supporting people like me? What can I do to make a difference?"

Any one of these questions would take many interviewers out of their comfort zone, but the NSU Board found the discussion to be an eye-opening, mind-expanding experience.

This kind of shift in perspective should not be limited to the classrooms of NSU. Organizations that wish to stay relevant need to be ready to answer these questions with honesty and transparency. The traditional interview process dedicated to vetting a candidate's experience and potential in a certain field is now being replaced with a two-way conversation, where the candidate is interviewing the organization as much as the other way around.

Just as job candidates prepare by recalling past experiences to share that will back up their skills, company interviewers also need to prepare for these valid questions. Human resource managers should have on hand specific examples, personal anecdotes, and hard numbers on the diversity of the company to back up their descriptions of the organization's progressive culture. The saying, "Don't just tell me, show me," can easily be applied here.

For me, the NSU exercise convincingly demonstrated the need for organizations to *focus less on impressing candidates and more on allowing candidates to express themselves.* That's not to say it isn't important to talk about your cutting-edge code of conduct or your commitment to valuing differences. But we also need to be sure we give all job seekers the safe space to ask heartfelt questions about their fears and aspirations—and then be prepared to answer them.

Another interesting and, perhaps, surprising challenge facing organizations is the intersection of the issues of diversity and inclusion with technology. In my experience, AI interconnects with D&I in a host of ways.

An article in *Science* magazine highlighted some of the potential issues, noting that research into the way computer tools such as language translation systems understand and use words reveals biases that reflect those of the human programmers who developed the tools. One study, for instance, found that programmers had unconsciously trained algorithms to link names and images of men with analytical and mechanical skills, while women were linked with childcare and homemaking chores. A commitment to D&I is essential to ensure that we are mitigating biases and minimizing inequities and inaccuracies in the AI tools we develop.

At my organization, we've created a diverse team of professionals to work on issues surrounding the connections between D&I and automation. Their goal is to ensure that automation promotes best-in-class D&I objectives. We've also involved more diverse groups of our people in tech-related service solutions. Technologists are joined on problem-solving teams by professionals with backgrounds in accounting, law, STEM subjects, and liberal arts. This approach reflects our belief, supported by practical experience, that diverse points of view yield better answers. The diversity we seek involves not only demographic traits like gender, age, and ethnicity but also cognitive diversity involving qualities like thinking, learning, and working styles.

Recognizing that we are still early in our journey, we are conscientious about acknowledging the "known unknown"—that threats of potential bias always lurk in the background. We strive to remember that technology is not neutral, and it's our responsibility to make it positive. Our experience offers up four crucial questions that business leaders should ask regarding our commitment to D&I:

- What must leaders do to ensure we stay alert to potential bias in policies, practices, and systems that may alienate some of our people and deprive us of the insights and knowledge they have to offer?
- What steps must leaders take to prevent machines from inheriting human bias and passing it along to future generations of technology tools?
- What role should leaders play in managing the relationship between people and changing technologies, particularly in establishing and maintaining trust, and in monitoring the long-term effects that technology may have on shaping human behaviors?
- How can leaders ensure we are actively listening to our professionals to gauge their sense of belonging and to understand what they need to remain fully engaged in the work of our organization?

My colleagues and I consider this next stage in our organizational evolution to be exciting and filled with opportunities. Being proactive and mindful of the intersection between automation and D&I as we progress along this journey will allow us to stay connected with one another as we reinforce our overall strategy—one that pursues success for the organization and its customers even as we keep our people front and center.

The Most Powerful Workplace Ingredient: Trust

FINALLY, ABOVE AND BEYOND the five key ingredients I've described as essential for companies that want to prepare for tomorrow's changes, there's another vital workplace element that is timeless yet ever-changing—a culture built on mutual trust.

For today's rising generations, trust is a huge issue in choosing and staying with an employer. It's also essential to driving a culture of innovation. Trust is to innovation what wings are to flight and air is to fire. Without trust, and the behavior it drives, where would we get the power, the spark to unlock innovation, or the confidence to try new things and ultimately grow?

The benefits of trust are tangible and real. Studies show that, on average, companies with proven cultures of trust register two to three times greater stock market returns, 50 percent lower employee turnover rates, and 50 percent higher-than-average levels of innovation, employee engagement, and customer satisfaction. According to the Edelman Trust Barometer, employees who trust their employer are far more likely to engage in beneficial actions on the organization's behalf and to advocate publicly for the organization.

Some fear that the fast and furious pace of new technology disruption is diminishing trust by depersonalizing day-to-day transactions and putting greater distance between people and their institutions. Is this true? Is trust really diminishing, or is a new meaning of trust evolving?

I think the reality is that the ways in which we built and maintained trust in the past are not always relevant in today's world. New technologies are transforming who, how, and what we trust, and challenging us to think differently about the very nature of trust. Just look at the rapid rise and mainstream acceptance of the sharing economy, which is completely built on trust, with peer-to-peer sharing of cars, bikes, houses, clothing, home goods, and even pets! Technology itself has enabled the development of these types of trust and made it possible to build entire industries upon them.

What can you and your organization do to build trust in a time of technological change and shifting perspectives on how trust itself is earned and fostered? Here are four recommendations:

- *Don't ignore generational factors.* Digital natives—Millennials and members of Gen X and Gen Z—readily accept advancing technologies and inherently trust them, while often distrusting more traditional institutions like brick-and-mortar banks and car dealerships. They grew up in a digital environment and have no qualms about fully exploiting technology's speed and ease of use. By contrast, digital immigrants like the boomers may be slower to adopt new transaction models. But they can be

brought onboard with supportive learning programs that provide a better understanding of the
time and cost value of new lifestyle conveniences.
Being aware of these generational differences and
addressing them in your messaging is critical to
building trust.

- *Harness a culture of trust as an innovation acceler-
ator.* Progressive leaders must create a trustworthy
workplace, including management and talent policies that enable employee creativity, promote agile
thinking and open expression of ideas, and, im-
portantly, minimize the fear of failure. Leaders
need to send a message from the top that tells employees they can safely assume a certain level of
risk as they seek to make positive change. Leaders
should also underscore the value of building connections with colleagues and building on one another's ideas and energy—another form of activity
that demands a basic level of interpersonal trust.

- *Remember that trust bonds employees, customers,
and stakeholders.* There are obvious benefits from
demonstrating to the public that your company's
products and services can be trusted—for example,
by showing them that adequate security and privacy safeguards are in place. Equally important is
ensuring that your own employees know about and
believe in your trust commitment. In their 2021
survey, the Edelman Trust Barometer researchers
found that, in today's fractured world, employers
have emerged as the most reliable source of infor-

mation for most workers. Companies can take advantage of this moment of opportunity by leading with purpose and providing consistently reliable information about how they are doing so.

- *Establish a corporate mission that embraces key global issues.* A company-wide commitment to socially responsible leadership in areas important to most of the world's population can help engage employees and build their trust in your role in both the business world and the broader world. Seek to develop policies and practices that demonstrate your organization's commitment to resolving environmental issues, reducing the impact of poverty, supporting human rights, and investing in continuous learning and wellness initiatives, among many others.

The best description of trust that I have heard is "economic gravity"—invisible yet impossible to ignore, crucial to maintaining order, and profoundly influential as a factor in market power. Our economy always has and always will run on trust, even as our relationship with trust changes with the times. With digitization of data and automation of processes, technology has given us greatly expanded abilities to assess and absorb information and to scale and grow the scope of our business activities. As a result, the significance of communicating and behaving in a trustworthy fashion has never been greater.

As technology shifts from working alongside us to working inside us, the need for a trusting relationship between technology and your employees has never been more important. If you build and harness trust as a key part of your organization's value system, you will unlock in your people the awesome power to innovate.

SPOTLIGHT

Leveling the Technological Playing Field

Throughout large parts of the world, significant groups of people are still excluded from the digital equation—left behind as more privileged populations find life becoming easier, more efficient, more prosperous, and better connected. For the poor in even the most developed nations—and especially for the most vulnerable in the emerging markets of the world—digital solutions to their problems seem as remote as they were 30 years ago, pre-smartphone, pre-internet, pre-artificial intelligence. The challenges of the COVID-19 pandemic opened an even wider window on how different and how challenging life can be when people do not have access to internet connections that facilitate learning, as well as digital technologies that can enable essential activities from monitoring water quality to receiving healthcare via virtual visits.

However, the story of global access to digital technologies is not entirely bleak. Even as we emerge from the harsh realities of the COVID era of 2020–21, studies by researchers at the World Economic Forum (WEF) point to examples of digital technologies being employed in ways that can help level the playing field and make life more livable for millions of those who have traditionally been underserved. Some developing countries with progressive goals have focused on leveraging the digital economy to human advantage. There are lessons here for the broader application of technology solutions to help preserve and extend human values and dignity.

One of the ways the developing nations are benefiting from digitization is through economic growth driven by information technology (IT) outsourcing. As a result, emerging markets in South America, Asia, and Africa are seeing a boom for local wage earners. Asia has now become the number-one region for providing IT outsourcing services to the rest of the world. In 2019, according to the Oxford Internet Institute, India was the world's largest supplier of online labor, with close to 24 percent of global freelance workers, followed by neighboring Bangladesh at 16 percent. While Indian freelancers dominate the areas of technology and software development, Bangladesh has become the top supplier of sales and marketing support services.

The Bangladeshi freelancing boom has created jobs in fields that include computer programming, web design, tax preparation, and search engine optimization, generating $100 million annually for some half a million people and offering a flexible new source of income that improves their lifestyles. Women particularly are benefiting. In the past, even highly educated Bangladeshi women had to sacrifice careers to stay home and care for children. Freelancing gives them new options to work from home, earning incomes in valuable foreign currencies.

Of course, there are plenty of challenges that Bangladesh still needs to meet in the years ahead. Most of the country, including the capital city of Dhaka, suffers from unreliable power sources that lead to frequent connection issues, fragmenting the mental focus needed for complex coding and service problem-solving. Slow broadband connections compound the problem. There currently is no easy system to facilitate foreign exchange payments, hindering receipt of income from overseas clients. And while the percentage of young Bangladeshis enjoying access to higher education has grown enormously, many still lack the skills needed to thrive in the global marketplace. The national government will need to focus on human capital development to help tens of millions of Bangladeshis to complete their journey to middle-class status.

Indonesia is another country that is at a critical juncture in the line between past and future. Between 2005 and 2020, the country made remarkable economic progress, reducing poverty levels to below 10 percent. Unfortunately, the COVID pandemic has threatened further progress. Lockdowns have closed schools and limited employment opportunities. To jumpstart future growth, young Indonesians need access to the right technology tools to equip them for 21st-century careers, and their families need financial support so they can stay in school rather than being forced into work at an early age.

WEF researchers point to programs that provide inclusive social financing, such as one supported by the Yayasan Cinta Anak Bangsa (YCAB) Foundation. The YCAB twin-track approach currently provides vocational training to more than four million youths in Indonesia and has supplied microloans to over 200,000 mothers with children in school. These small capital infusions—typically in amounts ranging between $100 and $200—help women to start and grow their own small businesses, giving them the financial resources to lift their families out of poverty and keep their children in school.

The impact of COVID-19 on education has been felt in countries around the world. WEF estimates that up to 1.6 billion learners were out of school at the peak of the pandemic, and as of early 2021 some 200 million remained without instruction, running the risk of damage to economic development for years to come. As with most economic and social problems, the poorest are the hardest hit. Some forms of training that offer the most immediate short-term benefit, such as vocational classes in fields like hair styling and electronic repair, have been devastated during the pandemic lockdowns, since they require hands-on, in-person learning.

On the positive side, the growth of online course offerings is enabling educators in some fields to reach more students at significantly lower cost per capita. Major global companies are getting involved, sponsoring programs in STEM learning and increasing exposure to these types of classes for girls. In Indonesia, the rise in online training is taking some forms that are culturally distinctive—for example, a hybrid hands-on / digital program that teaches batik, a process that uses dyes to make colorful fabrics sought after by consumers worldwide. The in-person classes have been held outdoors, so they have been largely uninterrupted by pandemic lockdowns, and students are being taught how to sell their batik goods on technology platforms like Instagram and Facebook.

Unfortunately, not all Indonesians have access to the online learning tools needed to supplement in-person learning. An estimated one-third of students in Indonesia lack access to such tools, along with one in four teachers. Making such access universal is one of the crucial challenges local leaders face in the decades to come.

Technological developments also have spurred a range of economic and social improvements in countries like India. By 2021, India boasted 624 million internet users, representing 45 percent of the population, while the number of mobile connections reached 1.1 billion, or 79 percent of the population. These trends are enabling Indian consumers to take advantage of new opportunities for education, business, and healthcare.

Other technologies are also contributing to a rising tide for many poor and disenfranchised communities. Data analysis tools, for example, are giving leaders in the developing world access to information about populations in remote areas who have often been omitted from traditional census reports or surveys. They are using big data and remote sensing techniques, such as drone aerial photography, long-wave infrared, and thermographic imaging, to extrapolate population information and improve the design of programs targeting social and economic development.

Digital solutions like these can help to ensure that the disadvantaged have the tools, methods, and platforms needed to fully participate in their communities and the local, national, and global economies. But making these resources available to everyone who needs them is a continuing challenge. Some of the big questions that government leaders, business executives, and policy makers need to address include the following:

- How can government-supported programs, for-profit ventures, and nonprofit initiatives be adroitly coordinated to make digital tools available not just to the middle class but to those still struggling to escape lives of poverty?

- How can leaders at the regional and local levels—including community members in countless remote villages in Asia, Africa, and Latin America—be empowered to help shape the use of digital technologies so that ordinary people benefit fully from them?
- How can projects to boost technology access in the developing world be designed to ensure that traditionally disfavored populations—including women, the disabled, and ethnic and religious minorities—are fully included?

Designing and implementing technology programs that will protect and enhance the full humanity of the hundreds of millions of people just entering the digital era will require both top-down initiatives from business and government leaders and bottom-up contributions from local community members, who need to feel a genuine sense of ownership and control over the ways these new technologies will reshape their lives in the decades to come.

CHAPTER SIX

The Path Back to Humanity

STEPPING BACK AND TAKING a clear-eyed view of where we are today with human-machine convergence yields signs we may soon come to a reckoning about the optimal path to augmenting humans to their maximum capability without stripping them of the most critical characteristics of humanity.

What are these critical characteristics? Philosophers, scholars, and scientists have wrestled with this question for thousands of years. I have no intention of wading into a complex intellectual challenge that our greatest thinkers have been unable to resolve. Instead, let's stay within the context of today's changing world, and simply identify some of the characteristics of human existence that I think we would all be unwilling to sacrifice in the journey toward a world of increasing convergence between humans and machines.

For this purpose, the characteristics I personally would hate to lose include powerful emotions such as love, hurt, wonder, pity, and fear; a sense of reverence for other living beings and for our planet; and intangible qualities that guide and shape the ways we behave, such as critical thinking, conscious choice, and the courage to do what we believe is right.

As we've seen, AI is now playing an increasingly important role as the decision-making function within autonomous systems. The result is a transformative moment in history, marked by the emergence of autonomous systems that are capable of controlling technological and organizational activities in manufacturing, logistics, security systems, and many other business processes.

These autonomous systems can mimic many of the activities we have long associated with being a thoughtful, analytical, and decisive human, including the ability to acquire or create data, information, ideas, and other forms of intellectual content; to interpret and develop knowledge using that content; to make decisions based on the content; and to act on those decisions. They have a growing capability to access data thanks to trends such as the proliferation of sensors in machines, tools, and apps, billions of which are connected into the emerging global network known as the Internet of Things (IoT). Further assisted by machine learning, autonomous systems can also program themselves by analyzing the content to which they have access, comparing the outcomes that follow from the decisions they make, and gradually improving their own decision-making processes in order to enhance those outcomes.

We see these capabilities being played out today in systems ranging from driverless cars and drones to systems for fraud detection, automated farm operation, loan origination, and high-speed, high-frequency financial trading. In the years to come, such autonomous systems will spread to countless other arenas, from transportation and construction to healthcare and scientific research—in fact, to practically every field of activity in which it's useful to gather and work with vast amounts of information.

These systems are capable of executing actions based on decision loops shaped by specific contexts and time frames. The duration of these decision loops varies enormously. The activities found in most hospital emergency rooms involve decision loops that last from a few seconds to a few minutes; typical medical diagnostics involve a few hours to a few days; and agricultural processes take a few weeks to a few months. By contrast, the control systems involved in high-frequency stock trading loops last for an average of 100 microseconds, while the systems used in driverless cars have decision loops that last an average of just a few milliseconds. (One microsecond is to a second what one second is to 11.5 days, and a millisecond is 1/1000th of a microsecond.) Such tiny increments of time seem nearly unfathomable to humans—yet autonomous systems are able to operate within this scale.

For many companies, their C-suite leaders, and their boards, it is still early days in the large-scale adoption of AI tools, machine learning, and autonomous systems. Other organizations are already caught up in the rush to implement these technologies in order to avoid being left behind by competitors. But undue haste can lead to thoughtless behaviors. Exceptional companies will examine their R&D expenditures and investments in AI and other technologies in order to enhance their competitive positions while also looking broadly at the implications of the new technologies for customers, employees, shareholders, communities where they operate, and other stakeholders.

The big question we all face: How can we implement the new autonomous systems without sacrificing or weakening the characteristics that define our humanity and that enable us to enjoy lives that are worth living? If we get the answer wrong—or fail to address the question in the first place—then the amazing capabilities of the technology will not benefit us in the long run.

Ways to Think About the Human Questions Related to the New Technologies

CORPORATE BOARDS ARE JUST BEGINNING to consider the potential impacts, positive and negative, of the new technologies on people and society. When these issues arise, Linda Goodspeed, a former chief information officer who sits on several boards of directors, likes to bring up two questions: "What decisions are we going to allow machines to make, and how are we going to audit those decisions?" Believing that thoughtful discussion about how to carefully adopt new technologies is critical to good governance, Goodspeed points to these questions as useful ways to launch the conversation.

Goodspeed's questions open the door to a series of other questions that business leaders need to ask to begin grappling with the deeper issues underlying the implementation of new technologies. Some of the questions are fact-based. If you are a board member, a C-suite executive, or any other kind of corporate leader, do you understand what all the functions in your organization are doing? Do you maintain an inventory of technological innovations at your organization? As you seek answers to these basic questions, you may discover that there are departments or divisions of your business that are already experimenting with or implementing new technologies in ways that could raise ethical issues or, at least, serious concerns about public perceptions. Company leaders should make sure they are staying abreast of new developments, practices, and behaviors that could, over time, subtly reshape the culture and even the very nature of their organizations.

Once you know the kinds of technological innovations your company is exploring, you can begin to figure out whether there should be guardrails in place to govern these innovations. If so, what are the fundamental values that should define those guardrails? Who will make the specific decisions regarding what the guardrails look like? How will the guardrails be implemented, monitored, and enforced?

To jumpstart this discussion about technology guardrails, here are some of the questions you may want to ask about any new decision-making technology, such as autonomous systems:

- *How will the new technology add value, quality, or efficiency* to our existing decision-making processes?
- *What are the tradeoffs we need to consider?* That is, in what ways might the new technology weaken

or undermine our efforts to produce value consistently for our business and for those we serve?

- *What kinds of safeguards have been incorporated into the new technology* to ensure that the interests of all those affected by it will be protected?

- *How will we ensure that business processes governed by the new technology remain faithful* to the regulatory, ethical, and quality frameworks within which we operate? Are they free of unintended biases?

- *How will the activities of the new technology be monitored?* Can decisions or activities be forensically examined if needed?

- *How will a regulatory body determine whether a new technological system is out of compliance?* How will the regulators determine what needs to be remediated? How will we perform that remediation and measure its effectiveness?

The importance of asking these questions is directly linked to the power and speed that make the new autonomous systems so potentially valuable. Whenever we deploy a business optimization process that is extremely complex and largely self-programming—and in that sense, more "intelligent" than the humans who designed it and specified its objectives—we run the risk of unpredictable and possibly irreversible consequences.

This is why auto companies have been testing driverless cars for years. The algorithms designed to ensure correct behaviors in responses to challenges like driving on the right side of the yellow line and stopping when there is another car or a red light in front of the vehicle operate on millisecond timescales. But it takes much longer for humans to study the responses of those algorithms in the vast array of varied real-world circumstances that may arise, and then to fine-tune the algorithms to address the dysfunctions that may be revealed.

There is an added complexity that is potentially more serious from the standpoint of human-machine convergence. The optimization process will demonstrate behaviors that could be classed either as ethical or unethical. This is true, in part, because existing data almost always reflects existing biases—which means that the decisions made by autonomous systems can replicate those biases. One classic example arose when an autonomous loan origination system was found to be arbitrarily denying loans to applicants from specific demographic groups. The system was informed by no deductive or scientific rule; the behavior arose from input inherent to the optimization process, yet we humans would call it unethical. The loan origination issue was years in the making and could take months or more to remediate. The expense of the remediation work and, perhaps more costly, the damage to the financial firm's brand would be significant.

Up until now, humans have always made decisions of this kind; technologies have always played a supporting role. Now, fewer humans are making decisions, and an autonomous system cannot be expected to answer the question "Why did you make that decision?"

The challenge we face is to rapidly develop the skills and competencies needed to define the risks associated with autonomous systems and to articulate and implement solutions for them. The situation resembles the one business faced in the mid-1990s when companies had to create from scratch a delivery capability for internet-based goods and services. They drew upon their previous experiences in strategy, architecture, testing, and operations, yet there was a significant element of circularity in their work; they ended up with the same premises and management tools that they started with, despite the fact that the entire paradigm had shifted and old ways of working had been disrupted. Some firms were able to adapt and flourish, while others were not. The same will be true of life with autonomous systems.

Ultimately, I believe the positives involved in use of autonomous systems will far outweigh the negatives. Nonetheless, the problems to be solved will be challenging, and for the next decade the choices we make will set precedents throughout global businesses. Organizations able to master and deploy workable, high-performing human-augmented solutions for this type of sophisticated technology will enjoy a powerful first-mover advantage.

I believe we can take action to help ensure that our answers tend to lead toward what most of us would consider our best future scenario. Doing so requires hewing to values and principles that can serve as foundations for defending and enhancing our humanity, in particular:

- Trust
- Ethics
- Sustainability
- Privacy
- Learning

It's true that these five values don't provide what we need to solve all problems in all contexts. They certainly wouldn't suffice to serve as guardrails against the dangers posed by the misuse of autonomous systems by those with extreme agendas, such as outright criminals or would-be dictators. But these values are among the most venerable yet modern lenses we can use to evaluate our decisions and actions. They reinforce the commitment to do no harm, which I believe most business leaders would willingly accept. They also provide a constructive framework for future innovation. They can help keep us on the right road, even if we sometimes briefly cross over into the wrong lane.

Using these five values as a guide will help ensure that humanity comes first while not hindering the pace and potential of technological advancement. The goal is not to limit innovation but to ensure that the focus is on benefiting humanity and society.

Trust

I'VE ALREADY CITED the definition of trust as a kind of "economic gravity." I really like that metaphor. Trust is fundamental to positive, constructive participation in the world of business. And beyond the business world, trust also serves as social gravity, educational gravity, even spiritual gravity. Our economy, our relationships, and our lives all run on trust.

But while gravity is a constant, the concept of trust is much more variable. Indeed, our relationship with trust has been reshaped with the times. In recent years, technology has obviously played a big role in this reshaping. We used to trust the restaurant reviews in the local paper; now we look to the restaurant's rankings on social media review sites. We used to trust the information reported by our favorite network news anchor; these days, our tendency toward confirmation bias leads many of us to rely on the Twitter star or podcast host whose views align most closely with our own.

As technology continues to be inextricably woven into the fabric of our lives, we'll need to consider a growing list of weighty questions about trust. For instance, which of the following would you trust today?

- An autonomously driven school bus to take your children to school
- An algorithm to make daily household purchases on your behalf
- An AI system to buy and sell stocks for your retirement savings account
- A machine learning tool to review your job performance and decide whether you are worthy of consideration for a promotion
- Facial recognition software to determine your innocence or guilt in a criminal case

Imagining one or more of these scenarios probably makes you feel anxious. Where will we draw the lines between acceptable and unacceptable uses of technology—and will those lines be drawn by us as individuals or by society as a whole?

Today, we're confronting gnarly issues around exploitation of social media platforms, unintended bias in AI and coding, and fears around the misuse of our personal data. We will continue to confront more of these types of issues as technology changes who, what, and how we trust.

Answering these complex questions is especially difficult now, when these technologies are still so new. It seems clear that building, earning, and fostering trust in this new environment requires a new approach. Here are three powerful steps that leaders can take to begin the process:

- *Cultivate a culture of trust inside our organizations.* Progressive leaders need to create trustworthy organizational environments that support critical and creative thinking. They need to lead by example, promoting agile thinking and encouraging honest and direct expression. A culture of trust makes everyone involved feel safer about expressing their values, facing up to their mistakes, and offering constructive solutions.

- *Demonstrate our own trustworthiness through concrete actions, not through promises.* Of course, you want to convince people that your group's ideas, products, and services can be trusted. But building trust is a classic case in which the rule of "Show, don't tell" must be applied. Suppose you've built a new smartphone app. Can you *show* consumers that you've taken the right steps to address trust-centered issues, such as quality control, security, and privacy? Suppose you're an employer in a tight labor market. Can you *show* your recruits and employees the seriousness of your commitment to trust?

- *Behave responsibly in regard to today's big ethical challenges.* Leaders who hope to earn the trust of the public need to engage issues of critical global importance, including environmental sustainability, poverty, human rights, and education. These issues are particularly on the radar of younger generations, such as Millennials and members of Gen Z, who are more likely to trust engaged leaders who are working toward making the world a better place.

Today's leaders must back up their words with actions that authentically represent their companies' values and link those values to policies, messaging, and programs designed to support employee involvement. Furthermore, organizational incentives, governance practices, and performance metrics must drive meaningful action in support of those same values. Without these concrete signs of commitment, all the words in the world won't earn the trust that companies need to thrive.

Ethics

DESPITE OUR BEST INTENTIONS, we humans sometimes make poor choices. Those poor choices often stem from our tendency to base our actions on intuition over facts—to choose what feels right rather than focusing on what's rational. That's not a big deal when the decision is whether to have that late-night snack or to sleep in on Saturday morning. But when the decisions are bigger, the potential to make bad or even unethical choices poses a heftier risk.

The convergence of humans and technology is a minefield of ethical issues. Our decisions around technology are often made with constructive optimism but without having been fully vetted. After all, it's easy to get caught up in the velocity and sheer momentum of change without pausing to thoroughly consider all the consequences. Have we fully weighed the good and bad of "social credit" systems that attempt to gauge citizens' behaviors, or the use of individual location tracking to monitor public health risks? The bright, shiny objects that captivate our attention and imagination can obscure their very real dangers.

That's why it's so important to build awareness of, and have genuine discussions about, what's truly at stake with these decisions. Encouraging people to take personal responsibility for their choices can add important perspective to these conversations. It also can help move us away from making decisions based on self-serving biases and toward decisions that are beneficial for society.

We need to have these discussions sooner rather than later. We will likely face challenging ethical crossroads over the next few years around such matters as machine learning and genetic engineering. What kinds of ethical decisions must we make when considering how—or even whether—we want to use machine-learning techniques that teach empathy to AI tools? How will we decide how far it's appropriate to go in using the new editing techniques to modify the human genome?

We must figure out how to effectively realize the full potential of AI and other technologies without sacrificing what it means to be human and what it means to be just and fair. We must be strategic about building our relationship with the machines inside us in much the same way we chose how to make the use of earlier technologies that worked with us.

Of course, the way businesses handled past waves of technological change was far from perfect. Consider, for example, the way society has dealt with the loss of jobs to automation. Being human, the managers of businesses sometimes committed errors in judgment and often made decisions based on the profit motive alone. Thoughtless applications of technology have produced some serious negative impacts, including economic disruption, job losses, and damage to the environment. Continued technological advancement has the potential to produce other harmful impacts—perhaps not as bad as the soot-filled skies of early industrial cities or the exploitation of child labor, but with serious economic reverberations and the risk that many people will potentially be left behind. All the more reason we need to ensure that people get the skills training and education they need to succeed in a digitized world, and that policy makers, academicians, and business leaders are alert to the ethical, economic, political, social, and environmental requirements of a society that values humanity.

Unfortunately, today's news all too often brings us stories about problems like technology addiction, the rise of cyberattacks and technology-based scams, and bad-faith decisions by tech companies. This is evidence of a weakened system.

People from all areas of society need to be better informed about the full range of impacts that may come from new technologies, not just the ways those technologies promise to make our lives easier. Those of us who want advancing technology to continue toward optimizing humanity, while also wanting to stay true to basic human ethical standards, must play a leading role in spreading this message and starting these conversations. We need to talk about the choices we face while reinforcing the importance of making wise, ethical decisions for the benefit of generations to come.

Sustainability

WE ALL NEED A HEALTHY, sustainable environment, including clean air, fresh water, and safe food. Beyond our natural world, there are other elements that contribute to a healthy environment, including human rights, ethical values, health, and well-being. These, too, are issues of sustainability, since a social and economic system that exploits individuals or groups and is manifestly unjust is built on a shaky foundation that invites upheaval.

In the last 30 years, the world has made tremendous efforts toward environmental sustainability. We're harnessing the power of technology to generate renewable energy from natural resources, such as the sun, the wind, and the tides. We've begun improving food-production capabilities to address health and hunger issues, and we've leveraged innovations to increase access to clean drinking water. Still, there is much work yet to be done.

Our collective efforts at global sustainability have reinforced an important truth: We are all connected, and what impacts one group will ultimately impact us all. Despite our rich diversity of people and cultures, we are inextricably bound together by our natural and human environment. With that as a given, how can we balance our world's need for sustainability with the desire to preserve and maximize humanity, especially as the role of government and public initiatives varies so widely across nations?

At this pivotal moment, it is no longer an option to ignore issues such as sustainability and climate change. Growth and competitiveness today require accountability by organizations and individuals alike. But what does that accountability look like? It starts with responsible business practices, including commitments to sustainable economic development and long-term value creation. In areas where government has faltered, businesses must do their part to support and protect their employees and their communities. In short, for-profit companies must "do well by doing good" when it comes to supporting the well-being of our natural and human environments.

So far, the record of business on this front is mixed. While we have seen some great strides from the private sector when it comes to conserving resources and averting climate change, the global business community at large has so far been unable to break its dependency on resource consumption to fuel economic growth. As of mid-2021, the Biden administration is incentivizing American utilities to go as carbon-free as possible by 2030. It has said it also will target the transportation industry to wean itself away from the use of diesel, gasoline, and oil. The success of such proposals largely depends on legislation supporting efficient mitigation mechanisms, such as a carbon tax and vehicle emission-reduction incentives. In emerging markets, where much of the world's economic growth is taking place, attitudes and regulations around environmental issues are even less mature than in developed markets.

In the spring of 2021, the International Monetary Fund (IMF) and the Organization for Economic Cooperation and Development (OECD) issued a report on tax policy and climate change, whose overarching finding was clear: "A progressive transition to net zero greenhouse gas emissions by around the middle of the century is essential for containing the risks of dangerous climate change."

The existing economic model, which includes a reliance on global supply chains, must evolve to continue improving economic development and quality of life while minimizing damage to the natural environment. Many put great hope in adoption of the *circular economy* model, which embodies the idea that economic systems should reduce their reliance on extraction of new materials and maximize reuse of existing materials and resources. The circular economy model reduces costs and adds business value by streamlining the supply chain. For instance, a company may agree to take products and materials back from consumers for recycling, or it may seek to lease products over and over rather than manufacturing and selling individual items.

One example: a Scandinavian company that makes clothes for children and expectant moms chose a subscription model that lets parents return items after their kids outgrow them. The result was a thriving business model that also kept tons of waste out of landfills. This way of doing business may appeal to businesses as a way to not only keep costs down but also to manage regulatory compliance, boost eligibility for tax breaks, and even burnish their brands as socially responsible corporate citizens.

Technology's continued convergence with humanity can help us pursue environmental sustainability. For instance, companies may use AI and sensors linked to the Internet of Things (IoT) to alert users when components of their gadgets and appliances are about to fail and assist them in replacing them, helping to lengthen product life cycles and mitigate the rapid acceleration of electronic waste around the world. AI and IoT can also lead to more efficient and sustainable agricultural practices that, in turn, will help address global food-insecurity issues. These advanced technologies are also making it easier to protect critical species and make better use of limited natural resources, such as water. In short, technology can be a true partner in our fight for a more sustainable world.

Another lever in this fight can be found in an unexpected place—Wall Street. Investors are increasingly holding companies accountable for their efforts to combat climate change as well as for their stances on social and governance issues. ESG (environmental, social, and governance) and CSR (corporate social responsibility) investing practices are fast gaining traction among retail and institutional investors. In a 2018 EY survey, 97 percent of respondents said they evaluate a target company's nonfinancial performance before investing. That demand has helped push corporate action to address social issues and led to increased efforts among companies to measure their progress.

What's more, companies are recognizing the bottom-line benefits of their environmental, social, and governance positions. Research shows that ESG efforts can lower operational costs, boost worker productivity, and enhance profitability. It also can reduce major risks: For instance, a company with a strong focus on environmental issues may realize it doesn't make sense to locate a new factory in a vulnerable coastal area that may be affected by rising sea levels. Proactive governance also can help companies avoid scandals that can cause major reputational damage.

ESG and CSR investment practices now being practiced by growing numbers of individuals and organizations will make a big difference in this pivotal moment for humanity. They are models for important change, and these collective efforts matter—incremental steps that can lead to a tipping point. Sustainable strategies are everyone's responsibility and should be factored into decisions big and small. They can help individuals, communities, and organizations grow and thrive while also helping to solve some of our greatest global challenges.

Privacy

PRIVACY IS AN ESSENTIAL INGREDIENT of our lives. We are thoughtful creatures and crave a certain amount of our own space. We all have varying points of view about privacy, but few would deny that privacy is a basic human right that we should have individual control over.

For all technology's benefits, it has inarguably had a negative impact on our personal privacy, nibbling away at the boundaries that define how we control our personal information. Of course, technology itself hasn't infringed on our privacy rights. Instead, the cause has been our willing participation in the technology built to share information, including some of our most personal data.

Social media, networking sites, apps for job hunting and dating—these are obvious examples of technology platforms we consciously opt into. But there also are hundreds if not thousands of information-gathering tools bundled into our use of retail websites, corporate sites, and other internet locations. Digital cookies track your every need and wish, and report back with yet more data on who you are and what you like. Then there's the growing number of data-hungry connected devices, such as smartphones, smart watches, and fitness trackers.

These sites, apps, and devices are just a few examples of how data about us are collected. There are also digital assistants in our kitchens, video doorbells that record every interaction at our front door, and the multitude of cameras dotting our landscape that track us throughout our days. With so much data being collected, technology has been specifically developed to more efficiently slice and dice all that information. AI-powered analytics can sift through enormous amounts of data with piercing efficiency, connecting dots to create a personal profile of your digital life that can inform how retailers sell to you, how political campaigns appeal to you, and how your insurance company sets your rates.

In years to come, the data game will only get more intense. Consider this future scenario: Your voice-activated digital assistant is being trained by AI to have more proactive interactions with you. Heading off to the movies? Your smart speaker may ask whether it can book you a rideshare service. Making a restaurant reservation? Your digital assistant might recall that you like a quiet table by a window and put in the special request for you. Developments like these will have privacy implications, from the data gathering itself to the way you may feel about sharing control of your life with a digital entity.

Privacy and the free flow of information are two sides of a double-edged sword. The same technologies that give us a wide array of innovations to connect and share with others also can be weaponized against us, often without our full awareness. For all these reasons, managing privacy in our data-hungry world requires a thoughtful approach that includes enforceable protections and the establishment of norms.

But putting up guardrails to protect our privacy will need to be done carefully. The firm boundaries needed to ensure individual privacy must also allow for the free flow of information that we are willing to share with friends, family members, and organizations whose intentions we trust. Strengthening laws to protect individual rights to privacy must be done without impinging on the other freedoms we cherish.

Learning

THE EARLY STAGES of the Fifth Industrial Revolution represent a period of great disruption and breathtaking innovations. They're laying the groundwork for a period of rich opportunity for those who are open-minded and eager to embrace the possibilities the future holds. Managing this period well requires that we think carefully about the technologies we use, but it is also about our own preparation for this new world we're co-creating with technology. What steps must we take in terms of education and work-life experiences that will help support technology's role in enhancing our lives?

Education is one of the most important paths to a better society. It plays a fundamental role in continuous innovation and in better understanding and mastering our lives and our world.

These days, education is a particularly pressing concern. Our job market is changing quickly as technology increasingly infiltrates our offices, boardrooms, and factory floors. The reality is that technology will soon be replacing significant numbers of both white- and blue-collar positions. In fact, technology is blurring the lines between these types of positions, creating a hybrid environment of what some call "no-collar" workers. As a result, education is moving into a critical state of flux, involving everything from how we educate our citizenry to who is responsible for that education.

The good news: Technologies like AI and virtual reality are beginning to transform how and where we learn. They're bringing new life to learning, offerings ways to combine words and images to create more immersive learning experiences. These powerful tools, which are increasingly cheap and easy to use, will provide greater access to educational opportunities. The future may bring everything from AI-driven tools that can help in the early identification and remediation of learning disabilities to personalized AI "teachers" that can guide individuals through lifelong education and job hunting.

Here are some of the ways our paths to education are changing in this new technological era:

- *Where we learn.* Technological developments have helped education break away from physical classrooms. Virtual tools are fueling alternative models for education and learning, increasing access and removing historic barriers of cost, time, and location that made it difficult for millions of people to get access to quality education. Many educational institutions are shifting from a mixture of in-person and online courses toward a goal of 80 percent virtual learning, spurred on in part by the COVID-19 pandemic. Why spend thousands of dollars and hours of commuting time to take a class when you can get the same education at low or no cost through a massive open online course (MOOC)? Meanwhile, "new collar" job programs offer specialized education to develop in-demand skills, and experiential programs offer a combination of cultural immersion, intellectual development, and character-building life skills. By extending these programs far beyond the traditional classroom, schools are leveling the educational playing field for everyone.

- *When we learn.* Today, education is no longer a once-and-done life stage. For productive members of the workforce, learning needs to be a dynamic state of being—a career-long, lifelong journey with many twists and turns, based on an ongoing commitment to acquiring knowledge, mastering fresh skills, and obtaining new experiences. That kind of education is what will help prepare workers and companies for the jobs of tomorrow. To meet the need, increasing numbers of organizations are playing a broader role in providing educational opportunities for their employees, investing more time and money in employee training and taking advantage of new technologies that make learning more accessible and affordable.

- *What we learn.* The growth of technology in the workplace is accelerating the need for some uniquely human skills. The trend toward science, technology, engineering and math (STEM) education in recent years has yielded great results. But now that effort needs to be balanced with a greater focus on essential business capabilities, such as effective teaming, listening, and problem-solving skills.

Employers increasingly need employees who can think, navigate, connect, and relate differently than in the past. They need employees who can act with empathy and resilience, understand 360-degree thinking and innovation, and willingly embrace disruption.

Upskilling at every level, beginning with postsecondary education and continuing through the careers of the most senior executives, will be a constant area of business focus, requiring frequently updated curricula and continually introducing new programs and tools—badging, bootcamps, undergraduate and graduate studies—all focused on the skills needed for today and the future. Some companies will find they can deliver high-quality learning more cost effectively than many formal educational institutions, also giving them the flexibility to pull in the best instructors and subject matter experts. Add to that the advantages of virtual-learning technologies, and a self-educating organization has the best of all worlds—a wide range of course options, excellent instructors, and an anytime/anywhere learning platform.

Facing the Challenges Together

MOST OF US ARE INTERESTED in being intentional about preserving uniquely human qualities. If we view the decisions we are about to face in transactional terms, I believe we have some idea of what constitutes a fair price to pay for the perceived benefits of technological enhancement, improved life quality, and greater longevity. But who will start the bidding? Who will negotiate the final offer? In short, how do we somehow all come together on making these crucial choices when there are so many stakeholders in our human ecosystem?

These are just a few of the key questions to start the honest debate we need to have:

- What aspects of our shared humanity must we commit to protecting and maximizing?
- Who is responsible for informing and educating people about the major issues society faces?
- How and where will the crucial policy debates of the coming years take place, and what can we do to ensure that everyone will have a voice?
- How can we increase the agency that individual humans have in framing their own futures?
- How can business, government, academia, and communities demonstrate their commitment to ethical policies and practices?
- What must companies do to ensure that they are focused on value creation for humanity as a whole as well as on maximizing their own profits?
- How can we ensure that government leaders make policy decisions driven more by human needs than by the quest for political power?

The transformation taking place is seismic, but the issues it raises are not always entirely clear. There is no single, unambiguous definition of "good" and "bad." The drive for advancement can blind us to unintended consequences, giving rise to new threats to humanity and society. Are we winning when we place an online order at nine A.M. and receive a delivery an hour later? Or are we losing if the electronic transaction jeopardizes the privacy of our personal data?

Somehow our disparate interests as separate entities—individuals, nationalities, businesses, governments, educational institutions, religious communities—will have to give way to a broader societal framework. There already are pockets of activity moving us in this direction. The corporate world is focused not only on efficiencies and productivity but on meeting urgent social needs. Government experts are working on ways to adjust social programs for a future of longer life spans and fewer jobs. Religious groups are exploring practical applications of age-old spiritual concepts through forms of action that range from meditation and mindfulness practices to social advocacy.

While darker forces may see opportunity in the technological ability to manipulate human communications and actions, the history of innovation suggests that, over the long haul, humankind will continue to follow the arc of progress. An array of forces—social, economic, political—will come together to help us get it right, with a few standout individuals helping to driving human progress through their powers of invention, their gift for collaboration, their commitment to consensus-building, and their sheer resilient leadership.

Reimagining the Human Lifespan— Technological Enablers and Social Implications

One major difference between humans and the rest of the animal kingdom is the fact that we understand from early in our lives that we will die. As a result, a meaningful percentage of our time living is spent thinking about, avoiding, or preparing for death. As poet Emily Dickinson wrote, "That it will never come again is what makes life so sweet." Which raises the question: What will happen if science and technology find ways to greatly extend human life, or, as some believe, even to "cure" death? Will we behave differently if we no longer face the certainty of death? What will happen to the great religions, most of which teach the notion that a desirable eternal life is the reward of an earthly life well lived?

There's no question that the human "expiration date" has been extended. The average lifespan of people born in the United States in 1900 was 20 to 25 years shorter than the lifespan of those born in 1950. The trend is continuing today: On average, those born after 1950 have added between one and two years to their life expectancy for every five-year interval since then. This means that if you were born during the 21st century, you have a 50 percent chance of living beyond your 100th birthday. And current developments in artificial intelligence, genetics, pharmaceuticals, and other scientific and medical areas have enormous potential to ratchet the process of life extension even further over the next few years.

It's possible that the first human to live to the age of 200 is alive somewhere on planet Earth today. Is this a thrilling possibility to imagine—or a disturbing one? Are we prepared—psychologically, socially, economically, and spiritually—to live rich and rewarding lives over a period that lasts 20, 30, or 50 percent longer than the lifespan our parents experienced?

There are risks and rewards inherent in our species' journey toward an even longer lifespan. The risks start with the ways that future life extension is most likely to be achieved. A number of wealthy and famous leaders from the worlds of business and technology have donated large sums to support research into anti-aging and life extension. On another front, about 20,000 people to date have invested their hopes in cryogenics, a process of quick-freezing a body soon after death in a bath of liquid nitrogen, hoping for restoration of life at a later point in time when the conditions that caused death have been cured.

Another version of the dream of immortality arises out of the so-called Singularity Movement. Members of this movement believe that artificial intelligence is about to reach an inflection point at which machines will be able to advance beyond human intelligence and ultimately become self-sufficient. The concept was first advanced in the 1950s by mathematician and engineer John von Neumann, who posited that "the ever-accelerating progress of technology . . . gives the appearance of approaching some essential singularity in the history of the race beyond which human affairs, as we know them, could not continue." Von Neumann also defined "the singularity" as the moment beyond which "technological progress will become incomprehensively rapid and complicated."

More recently, writers like futurist and technologist Ray Kurzweil have posited that the Singularity will feature the introduction of technology directly into the human body for the purpose of empowering and sustaining life—a logical extension of the human-machine convergence we've discussed in this book. Kurzweil believes that fantastically extended human longevity will be achieved when nanobots course through our blood streams, enhancing our physical capabilities by making us more machine than human and therefore less vulnerable to disease, aging, and death.

Perhaps the most promising route to a longer, improved human lifespan is the evolving science of epigenetics, which analyzes how behavioral and environmental factors can affect the workings of our genes. Like so many light switches, epigenetic changes modify gene activity without changing the underlying sequence of DNA molecules.

Epigenetic switching mechanisms can be seen at work in the natural course of life. When we get a cut, for example, the genes for making new skin cells are switched on. Similarly, when we gain a new skill, the expression of genetic factors in our brain is modified. Even our patterns of diet, sleep, and exercise can cause chemical modifications that impact the way our genetic inheritance is expressed.

Thus, epigenetics does not change our genetic makeup, but it enables changes in how one's body reads and carries out the commands embedding in its DNA sequence. Epigenetics helps to explain how identical twins can have the same DNA but demonstrate different talents, interests, and physical attributes.

Scientists working in applied epigenetics are seeking ways to manipulate these "light switches" as they relate to serious illnesses and other life-limiting conditions, from obesity to Alzheimer's disease and cancer. One example is CRISPR gene-editing technology, which is named after a specific family of DNA sequences that can be modified and used to make changes in genes within living cells. In effect, CRISPR technology can be used to find a bit of DNA inside a cell and turn it on or off without impacting the DNA sequence itself. It has the potential to transform medicine, enhance food products, defend against viruses, and perhaps help cure serious illnesses.

Epigenetics most recently played an important role in tests and vaccinations developed to detect and fight COVID-19. According to a study by Weill Cornell Medicine, the virus drives a distinctive reprogramming of gene activity in an infected patient's immune cells. Epigenetics helped identify the distinctive signature of the virus and determine ways to target treatments for it. It's possible that epigenetic tools may be able not merely to prolong human life but also to maintain and improve the quality of life even over vastly extended periods of time—an essential requirement if a 100-, 120-, or 150-year lifespan is to be a rewarding experience rather than a painful burden.

Even so, the social, economic, and political implications of a much longer average lifespan are difficult to comprehend. Virtually every aspect of our lives would change. The questions we'll need to grapple with are numerous:

- What will happen to workers who experience careers that span 70 years rather than 40? Will they be able to keep up with the new skills needed in a continually evolving workplace?

- How will relationships among work teams be affected when the generation gap between the youngest team members and the most veteran is far greater than it is today?
- How will younger workers react when opportunities for promotion are limited by older workers who remain fixed in high-ranking jobs for many extra years?
- What will happen to individuals with physically demanding, stressful jobs—firefighters, construction workers, emergency room nurses—if their lifespans are extended far beyond their ability to remain at work? Will it be possible to reform programs like Social Security to support such individuals for forty, fifty, or sixty years beyond retirement age?
- How must social expectations, values, and attitudes evolve so that the "super elderly" will be able to contribute and participate positively in the life of their communities and the nation?

The answers to questions like these will take time to work out. But it seems clear that developing a human-centered approach to the longer lifespans of the future will require an ecosystem response—one in which businesses, governments, nonprofit organizations, community organizations, and individuals all work together to design, support, and maintain the infrastructure needed to allow millions of super-elderly citizens live longer lives that are fulfilling, healthy, and satisfying.

Conclusion: Observations of a Transformation Optimist

ONE OF MY GREATEST CHALLENGES in writing this book has been trying to keep up with the unrelenting evolution of technology developments that are impacting and even reshaping humanity. Eventually I recognized the irony of the situation: The incredible difficulty of staying on top of change underscored the importance of thinking clearly about where this change is taking us.

There was a certain relief in this realization. I didn't have to keep up, I just needed to speak up—not to oppose technological progress, but to make a case for being as thoughtful as possible about what it means for our future.

As a transformation optimist, I want to see the world and our society as it should be, not merely as it can be. I want to see technological transformation and the fusion of people and technology empowering humans to be both brilliant and better, more innovative and more enlightened. Hence the themes that run through these chapters: leading with purpose; putting people first; focusing on the art and science of the possible; maintaining

Historic church near the Great Sacandaga Lake, New York
Photo by Martin Fiore

steadfast values around trust, ethics, diversity, and sustainability; protecting personal privacy; promoting a lifelong sense of wonder and a commitment to continuous learning; and never ceasing to ask the important questions.

Because we are at an inflection point in the arc of human history, we need safe spaces in which to wrestle with the deep questions about where we go from here. We know already some of what's next. The world continues to globalize. The ability to innovate is steadily being democratized, empowering millions of people equipped with smart devices connected to ever-stronger digital networks rather than being restricted to a relative few. The rising generations are seriously focused on addressing many of the big issues that threaten our planet and the quality of human life, with young people like the Parkland School shooting survivors, activist poet Amanda Gorman, and environmental crusader Greta Thunberg as just three examples.

These trends combine to make today an ideal time to reimagine a future for humanity in which longer, healthier lives of increased creativity can be enjoyed in a world blessed with sustainable environmental quality. To make this happen, a global ecosystem of diverse partners must be created. Let's bring together individuals from many walks of life as well as leaders from business, academia, government, nonprofit organizations, and local communities, ensuring space at the table for people representing all dimensions of diversity, so that everyone has a voice in going forward. Navigating a path that draws upon all these various points of view and priorities will be challenging but also constructive. We'll arrive at the best possible way forward by raising a broad range of questions and testing solutions that work for many constituencies.

There's a powerful saying often attributed to Antoine de Saint-Exupery, the French aviator and author: "If you want to build a ship, don't drum up people together to collect wood and don't assign them tasks and work, but rather teach them to long for the endless immensity of the sea." The vessel we need to build is a vision of the future focused on betterment of the human condition and protection of what is unassailably essential about human life. But it must be a goal shared and cultivated by many if we hope to build a craft that is sturdy and capacious enough to carry all of humankind on the grand journey before us.

Striking the proper balance between the rights and desires of the individual and the needs of society will always be challenging. The ecosystem approach can help us make connections to one another as parts of something bigger—communities in which we work, learn, worship, legislate, create, and contribute together, often connected through more than one spoke on the ecosystem wheel. Our mission is to engage and build relationships with those who share our yearning for a long-lasting, high-quality life for the greatest possible number of people.

Some may say that a project like this can't be started until formal institutions like governments, corporations, and legal systems commit their resources and influence to support it. But history has taught us that one, 10, 100, or 1,000 individuals can make a difference. We've seen powerful protest movements launched based on a lone parent's impassioned plea for justice; we've seen how a single person or a small group can leverage platforms like social media and tools like crowdsourcing to achieve amazing results.

I don't pretend to have all the answers, but I hope I've given some context to how we got to this moment and fostered some useful discussions about where we should go from here. Tech innovations often seem to take on a life of their own, but they are created by people and answerable to the needs of humankind. We can all play a role, whether as dreamers, developers, enablers, or users.

While it is true that "hope is not a strategy," I do believe that every successful strategy is underpinned and inspired by hope. Let's commit to a strategy that enables humanity to be reimagined, supported, and inspired by hope, and dedicated to changes that result in the greatest benefit to humanity.

Acknowledgments

I AM DEEPLY GRATEFUL to the colleagues, business associates, academic partners, students, family, and friends who inspired and helped with this book. Thanks for all the ideas, insights, and excellent questions that came from numerous meetings, panel discussions, Q&A sessions, and daily conversations, each one informing the point of view expressed in these pages.

In addition, this book would not have been possible without the dedication and passion of Pam Middleton, who helped turn this idea into a reality. I also want to express special thanks to Karl Weber and his able team at Rivertowns Books for their skills and expertise in guiding our way through this process. I also appreciate the work of Abby Trexler, Brooke Ferreri, and Ally O'Hara at Hot Paper Lantern for their help in giving voice to our messages.

Along the way there also were idea contributions, reviews of material, and moral support from friends, family, and colleagues who generously gave of their personal time and energy, including Bill Campbell, Michele Geist, Meggan Moore, Doreen Sullivan, and Esther Dijkdrenth. My thanks to them as well.

M.F.

Source Notes

Page 14, *One remarkable recent example:* Borislav Marinov, "The Number of Companies Making Industrial Exoskeletons Has Been Quietly Increasing for the Past Five Years," *Forbes,* September 24, 2020, https://www.forbes.com/sites/bslavmarinov/2020/09/24/the-number-of-companies-making-industrial-exoskeletons-has-been-quietly-increasing-for-the-past-five-years/?sh=40981dec7bf4.

Page 17, *As the example of RPA suggests:* McKinsey Global Institute, *A Future That Works: Employment, Automation, and Productivity,* McKinsey & Company, January 2017, https://www.mckinsey.com/~/media/mckinsey/featured%20insights/Digital%20Disruption/Harnessing%20automation%20for%20a%20future%20that%20works/MGI-A-future-that-works-Executive-summary.ashx.

Page 18, *However, process automation did increase:* Grand View Research, "Robotic Process Automation Market Size,

Share & Trends Analysis Report By Type, By Service, By Application, By Deployment, By Organization, By Region, and Segment Forecasts, 2021-2028," April 2021, https://www.grandviewresearch.com/industry-analysis/robotic-process-automation-rpa-market.

Page 20, *Sales of plug-in electric vehicles:* Mark Kane, "More than 3 million electric plug-in cars were sold in 2020," *Inside EVs,* February 2, 2021, https://insideevs.com/news/485298/global-plugin-car-sales-december-2020/.

Page 22, *There may be some prophecy:* John Howard, Vladimir Murashov, Brian D. Lowe, and Jack Lu, "Industrial Eoskeletons," *NIOSH Science Blog,* January 7, 2020, https://blogs.cdc.gov/niosh-science-blog/2020/01/07/industrial-exoskeletons/.

Page 26, *In April 2021, it was reported:* Emojipedia on Twitter, April 1, 2021, https://titter.com/emojipedia/status/1377614192681836553?lang=en.

Page 26, *In mid-July 2021:* Steve Case, "Innovation Moves to Middle America," *Wall Street Journal,* July 13, 2021, https://www.wsj.com/articles/innovation-moves-to-middle-america-11626199747.

Page 27, *Recommendations from experts at UNESCO:* UNESCO Global Education Coalition, *Supporting learning recovery one year into COVID-19: the Global Education Coalition in Action,* UNESCO website, https://unesdoc.unesco.org/ark:/48223/pf0000376061.

Page 28, *One study estimates:* Susan Lund, Any Madgavkar, James Manyika, and Sven Smit, *What's next for remote work:*

An analysis of 2,000 tasks, 800 jobs, and nine countries, McKjnsey Global Institute, November 23, 2020, https://www.mckinsey.com/featured-insights/future-of-work/whats-next-for-remote-work-an-analysis-of-2000-tasks-800-jobs-and-nine-countries#.

Page 28, *New York City reported:* Robert Frank, "Manhattan rental market plunges, leaving 15,000 empty apartments in Agust," CNBC, September 10, 2020, https://www.cnbc.com/2020/09/10/manhattan-rental-market-plunges-leaving-15000-empty-apartments.html.

Page 30, *In the face of this challenge:* "Jerry Kaplan on AI, Rbots and Society: Future Will Be 'More Like Star Trek Than Terminator,'" *Baidu USA,* November 19, 2015, https://usa.baidu.com/page/2/?search=2064763.

Page 33, *Cultural analyst Sherry Turkle:* See, for example, Ldsay Holmes, "Sherry Turkle on How Technology Can Impact Human Conversation," *HuffPost,* April 21, 2016, https://www.huffpost.com/entry/sherry-turkle-pioneers_n_5716699be4b06f35cb70c571.

Page 38, *Scientists estimate:* Rupert R. A. Bourne *et al.,* "Trends in prevalence of blindness and near vision impairment over 30 years: an analysis for the Global Burden of Disease study," *The Lancet,* December 1, 2020, https://www.thelancet.com/action/showPdf?pii=S2214-109X%2820%2930425-3.

Page 38, *This is the focus of the work:* Fabienne Lang, "New Implant for the Blind Links Directly into the Brain," *Interesting Engineering,* February 7, 2020, https://interestingengineering.com/new-implant-for-the-blind-links-directly-into-the-brain.

Page 41, *In July 2021, The New England Journal of Medicine:*
Pam Belluck, "Tapping into the Brain to Help a Paralyzed Man
Speak," *New York Times,* July 14, 2021,
https://www.ntimes.com/2021/07/14/health/speech-brain-
iplant-cmputer.html.

Page 45, *In 2000, the International Monetary Fund:* IMF Staff,
Globalization: Threat or Opportunity? April 12, 2000,
https://www.imf.org/external/np/exr/ib/2000/041200to.htm.

Page 50: *In fact, leading economists have credited:* See, for example, Richard Anderson and Kevin L. Kliesen, "The 1990s Acceleration in Labor Productivity: Causes and Measurement,"
Federal Reserve Bank of St. Louis Review, May 2006,
https://www.researchgate.net/publica-
tion/5047431_The_1990s_Aation_in_Labor_Productiv-
ity_Causes_and_Measurement.

Page 51, *One startup born at the Massachusetts Institute of
Technology:* Will Knight, "This Chip for AI Works Using Light,
Not Electrons," *Wired,* March 10, 2021,
https://www.wired.com/story/chip-ai-works-using-light-not-
electrons/.

Page 57, *There's reason to believe the rest of the world: Global
Study on Homicide,* United Nations Office on Drugs and Crime,
2019, https://www.unodc.org/documents/data-and-analy-
sis/gsh/Booklet1.pdf.

Page 57, *According to the World Economic Forum:* Ceri Parker,
"What if we get things right? Visions for 2030," World Economic Forum, October 29, 2019, https://www.wefo-
rum.org/agenda/2019/10/future-predictions-what-if-get-things-
right-visions-for-2030/.

Page 58, *In 2019, total winnings:* Jonathan Shieber, "Fortnite World Cup has handed out $30 million in prizes, and cemented its spot in the culture," *TechCrunch,* July 28, 2019, https://techcrunch.com/2019/07/28/fortnite-world-cup-has-handed-out-30-million-in-prizes-and-cemented-its-spot-in-the-culture/.

Page 59, *AI traffic-management tools that use sensors:* Henry Williams, "Artificial Intelligence May Make Traffic Congestion a Thing of the Past," *Wall Street Journal,* June 26, 2018, https://www.wsj.com/articles/artificial-intelligence-may-make-traffic-congestion-a-thing-of-the-past-1530043151.

Page 59, *With more than 90 percent of urban areas:* Robert Muggah, "The world's coastal cities are going under. Here's how some are fighting back," World Economic Forum, January 16, 2019, https://www.weforum.org/agenda/2019/01/the-world-s-coastal-cities-are-going-under-here-is-how-some-are-fighting-back/.

Page 59, *Another 1.6 billion people:* Daniel Cusick, "Cities Pledge More Green Space to Combat Urban Heat," E&E News, *Scientific American,* July 20, 2021, https://www.scientificameri-can.com/article/cities-pledge-more-green-space-to-combat-urban-heat/.

Page 61, *More than half of the world's population:* Iman Ghosh, "70 years of urban growth in 1 infographic," World Economic Forum, September 3, 2019, https://www.wefo-rum.org/agenda/2019/09/mapped-the-dramatic-global-rise-of-urbanization-1950-2020/.

Page 62, *As just one example of the challenges we face:* "Ensuring Prosperity in a Water-stressed World," World Resources

Institute website, accessed June 7, 2021,
https://www.wri.org/water.

Page 62, *New Clark City in the Philippines:* Olivia Rosane,
"Philippines Plans Manhattan-Sized Green City," *EcoWatch,*
May 14, 2018, https://www.ecowatch.com/philippines-green-
city-2568759284.html.

Page 63, *From the private sector perspective: Better Business,
Better World: The Report of the Business & Sustainable Development Commission,* January 2017, https://sustainabledevelop-
ment.un.org/content/documents/2399BetterBusinessBetterWor
ld.pdf.

Page 63, *Providing further underpinning:* "Business Roundtable
Redefines the Purpose of a Corporation to Promote 'An Economy That Serves All Americans,'" Business Roundtable, August
19, 2019, https://www.businessroundtable.org/business-
roundtable-redefines-the-purpose-of-a-corporation-to-
pomote-an-economy-that-serves-all-americans.

Page 63, *The UN estimates that in 2019:* "The number of international migrants reaches 272 million, continuing an upward
trend in all world regions, says UN," United Nations Department of Economic and Social Affairs, September 17, 2017,
https://www.un.org/development/desa/en/news/population/in-
ternational-migrant-stock-2019.html.

Page 64, *And in 2018, for the first time ever:* Hannah Ritchie,
"The world population is changing: For the first time there are
more people over 64 than children younger than 5," Our World
in Data, May 23, 2019, https://ourworldindata.org/population-
aged-65-outnumber-children.

Page 66, *As historian Ada Palmer describes it:* For an interesting summary of this history, see Dave Roos, "7 Ways the Printing Press Changed the World," *History Stories,* September 3, 2019, https://www.history.com/news/printing-press-renaissance.

Page 68, *As Marshall McLuhan wrote in his book:* Marshall McLuhan, *The Gutenberg Galaxy* (Toronto: University of Toronto Press, 1962).

Page 68, *Standing at the edge:* Sven Birkerts, *The Gutenberg Eegies: The Fate of Reading in an Electronic Age* (New York: Farrar, Straus and Giroux, 2006).

Page 69, *Childhood mortality rates:* Max Roser, Hannah Ritchie, and Bernadeta Dadonaite, "Child and Infant Mortality," Our World in Data, https://ourworldindata.org/child-mortality.

Page 69, *Over the past 50 years:* "GDP (current US$," World Bank Data, https://data.worldbank.org/indicator/NY.GDP.MKTP.CD.

Page 70, *Also in just the past 25 years:* "Decline of Global Etreme Poverty Continues but Has Slowed: World Bank," World Bank press release, September 19, 2018, https://www.worldbank.org/en/news/press-release/2018/09/19/decline-of-global-extreme-poverty-cntinues-but-has-slowed-world-bank.

Page 73, *The Varkey Foundation, a UK-based nonprofit:* Emma Broadbent *et al., Generation Z: Global Citizenship Survey,* Varkey Foundation, January 2017, https://www.vrkeyfoundation.org/media/4487/global-young-people-report-single-pages-new.pdf.

Page 77, *As a result, telemedicine is expected to grow:* "Tele-medicine market—growth, trends, COVID-19 impact, and fore-casts (2021-2026)," Mordor Intelligence, https://www.mordorintelligence.com/industry-reports/global-telemedicine-market-industry.

Page 84, *Today such machines are commonplace:* Christopher Mims, "On the 100th Anniversary of 'Robot,' They're Finally Taking Over," *Wall Street Journal,* January 23, 2021, https://www.wsj.com/articles/on-the-100th-anniversary-of-rbot-theyre-finally-taking-over-11611378002.

Page 85, *Research by experts at Stanford University:* Lynda Flowers *et al.,* "Medicare Spends More on Socially Isolated Adults," AARP Public Policy Institute, November 27, 2017, https://www.aarp.org/ppi/info-2017/medicare-spends-more-on-socially-isolated-older-adults.html.

Page 87, *Consumers are engines:* Kai Ryssdal and Maria Hnhorst, "What's Gonna Happen to the Consumer Economy?" *Marketplace,* April 6, 2020, https://www.mket-place.org/2020/04/06/whats-gonna-happen-to-the-consumer-economy/.

Page 87, *And they want more than just great products:* "State of the Connected Customer," Salesforce, 2019, https://c1.sfdc-static.com/content/dam/web/en_us/www/as-sets/pdf/salesforce-state-of-the-connected-customer-report-2019.pdf.

Page 88, *One, a patent for what it calls:* Anna Rose Welch, "Amazon Patents 'Anticipatory Shipping,'" *Retail IT,* January 23, 2014, https://www.retailitinsights.com/doc/amazon-patents-an-ticipatory-shipping-0001.

Page 88, *The second patent is for offering:* Greg Bensinger, "When Drones Aren't Enough, Amazon Envisions Delivery Trucks with 3D Printers," *Wall Street Journal,* February 26, 2015, https://www.wsj.com/articles/BL-DGB-40587.

Page 91, *A report by the World Economic Forum: The Future of Financial Services: How disruptive innovations are reshaping the way financial services are structured, provisioned and consumed,* World Economic Forum, June 2015, http://www3.weforum.org/docs/WEF_The_future__of_financial_services.pdf.

Page 101, *An EY study in 2019:* John de Yonge, "For CEOs, are the days of sidelining global challenges numbered?" Ernst & Young, July 8, 2019, https://www.ey.com/en_us/growth/ceo-imperative-global-challenges.

Page 103, *A 2019 study by Cone:* "Americans Willing To Buy Or Boycott Companies Based On Corporate Values, According To New Research By Cone Communications," Cone Communications, May 17, 2017, https://www.conecomm.com/news-blog/2017/5/15/americans-willing-to-buy-or-boycott-companies-based-on-corporate-values-according-to-new-research-by-cone-communications.

Page 103, *A 2019 study by Cone of Gen Z attitudes:* "90 Percent of Gen Z Tired of How Negative and Divided Our Country is Around Important Issues, According to Research by Porter Novelli/Cone," Cone Communications, October 23, 2019, https://www.conecomm.com/news-blog/2019/10/22/90-percent-of-gen-z-tired-of-how-negative-and-divided-our-country-is-around-important-issues-according-to-research-by-porter-novellicone.

Page 109, *One example: Freight Farms:* Freight Farms, https://www.freightfarms.com.

Page 110, *As a result, the city contains:* John Gallagher, "Social media is arguing about how much vacant land is in Detroit— and the number matters," *Detroit Free Press,* October 26, 2019, https://www.freep.com/story/money/business/john-gal-lagher/2019/10/26/detroit-vacant-land/4056467002/.

Page 111, *It's estimated that the total land mass:* Elijah Chiland, "In LA, land dedicated to parking is larger than Manhattan," *Curbed Los Angeles,* November 30, 2018, https://la.curbed.com/2018/11/30/18119646/los-angeles-parking-lots-total-size-development.

Page 112, *A 2014 article in The Social and Economic Review:* Kubi Ackerman *et al.,* "Sustainable Food Systems for Future Cities: The Potential of Urban Agriculture," *Economic and Social Review,* Vol. 45, No. 2, summer 2014, pages 189-206.

Page 112, *As of early 2021, hundreds of construction projects:* Aria Bendix, "These 3D-printed homes can be built for less than $4,000 in just 24 hours," *Business Insider,* March 12, 2019, https://www.businessinsider.com/3d-homes-that-take-24-hours-and-less-than-4000-to-print-2018-9.

Page 115, *Jim Clifton, former CEO:* Jim Clifton, *The Coming Jobs War* (Washington, D.C.: Gallup Press, 2013).

Page 118, *A research report published in March 2021:* McKinsey Global Institute, "Will productivity and growth return after the Covid-19 crisis?" McKinsey & Company, March 30, 2021, https://www.mckinsey.com/industries/public-and-social-sec-tor/our-insights/will-productivity-and-growth-return-after-the-covid-19-crisis.

Page 119, *A restaurant owner in California:* Will Knight, "Serve Food in Far-Away Restaurants—Straight from Your Couch," *Wired,* April 7, 2021, https://www.wired.com/story/serve-food-restaurants-from-couch/.

Page 119, *Managers of a meat processing plant:* Will Knight, "Covid Brings Automation to the Workplace, Killing Some Jobs," *Wired,* June 7, 2021, https://www.wired.com/story/covid-brings-automation-workplace-killing-some-jobs/.

Page 120, *A fast-food chain in Ohio: Ibid.*

Page 124, *For example, a 2017 report co-sponsored by IBM: The Quant Crunch,* Burning Glass Technologies, IBM, and the Business Higher Education Forum, 2017, https://www.ibm.com/downloads/cas/3RL3VXGA.

Page 124, *By early 2020, the Dice Tech Job Report: Dice Tech Job Report: The Fastest Growing Hubs, Roles and Skills,* Dice, http://mketing.dice.com/pdf/2020/Dice_2020_Tech_Job_Report.pdf.

Page 127: *Evidence is provided by these stats:* "Women in Sience, Technology, Engineering, and Mathematics (Quick Take)," Catalyst website, August 4, 2020, https://www.catalyst.org/research/women-in-science-technology-engineering-and-mathematics-stem/.

Page 128, *Dana Suskind is a practicing surgeon:* Emily Macmillan and Sara Serritella, "Thirty Million Words research reaches parents through book, programming," *UChicago News,* November 4, 2015, https://news.uchicago.edu/story/thirty-million-words-research-reaches-parents-through-book-programming.

Page 128, *Gregg Renfrew is an entrepreneur:* David Gelles, "Gregg Renfrew of Beautycounter on Toxic Chemicals and Getting Fired by Messenger," *New York Times,* November 21, 2018, https://www.nytimes.com/2018/11/21/business/gregg-renfrew-beautycounter-corner-office.html.

Page 129: *Marsha Lovett is deeply involved:* Marsha Lovett, "An Exciting Time to Be an Educator," *Digital Thistle,* http://www.digitalthistle.org/head-school/exciting-time-educator/.

Page 129, *There are many other women:* "Fei-Fei Li," Stanford Profiles, https://profiles.stanford.edu/fei-fei-li; Sofia Lotto Persio, "Inspiring Women to Become Sustainability Leaders in Engineering: The Yewande Akinola Vision," *Forbes,* March 6, 2021, https://www.forbes.com/sites/sofialottopersio/2021/03/06/inspiring-women-to-become-sustainability-leaders-in-eineering-the-yewande-akinola-vsion/?sh=c9ec68710ca7; Lisa M. Jarvis, "A Day with Jennifer Doudna: Trying to Keep Up with One of the World's Most Sought-After Scientists," *C&EN,* March 8, 2020, ttps://cen.acs.org/biological-chemistry/gene-editing/A-day-with-Jennifer-Doudna-Trying-to-keep-up-with-one-of-the-world-most-sought-after-scientists/98/i9.

Page 133, *For example, LinkedIn reported in 2020:* Deana (Lazzaroni) Pate, "The Top Skills Companies Need Most in 2020—And How to Learn Them," LinkedIn Learning Solutions, January 13, 2020, https://www.linkedin.com/business/learning/blog/top-skills-and-courses/the-skills-companies-need-most-in-2020and-how-to-learn-them.

Page 134, *The Graduate Management Admission Council's annual: Demand of Graduate Management Talent: 2021 Hiring*

Projections and Salary Trends, Graduate Management Admission Council, June 2021, https://www.gmac.com/-/media/files/gmac/research/employment-outlook/2021_crs-demand-of-gm-talent.pdf.

Page 134, *The Wall Street Journal has further highlighted:* Douglas Belkin, "Exclusive Test Data: Many Colleges Fail to Improve Critical-Thinking Skills," *Wall Street Journal,* June 5, 2017, https://www.wsj.com/articles/exclusive-test-data-many-colleges-fail-to-improve-critical-thinking-skills-1496686662.

Page 135, *In his book Deep Work:* Cal Newport, *Deep Work: Rules for Focused Success in a Distracted World* (New York: Grand Central, 2016).

Page 136, *In advocating for more young people:* J. M. Olejarz, "Liberal Arts in the Data Age," *Harvard Business Review,* July-August 2017.

Page 136, *One educational psychologist defines critical thinking:* Richard W. Paul, "Critical Thinking: Basic Questions & Answers," Foundation for Critical Thinking, https://www.criticalthinking.org/pages/critical-thinking-basic-questions-amp-answers/409.

Page 138, *According to the Department of Labor:* "Data and Statistics: Registered Apprenticeship National Results, Fiscal Year 2020," U.S. Department of Labor, https://www.dol.gov/agencies/eta/apprenticeship/about/statistics/2020.

Page 139, *For those interested in a more significant commitment:* "Apprenticeship Job Finder," U.S. Department of Labor website, https://www.apprenticeship.gov/apprenticeship-job-finder.

Page 141, *"Each time one skill becomes automatic"*: Robert Greene, *Mastery* (New York: Penguin, 2012).

Page 141, *Extreme Ownership is a book on leadership:* Jocko Willink and Leif Babin, *Extreme Ownership: How U.S. Navy Seals Lead and Win* (Sydney: Macmillan Australia, 2018).

Page 143, *Some experts in human biology:* Gerard Taylor, "Scientist thinks the world's first 200-year-old person has already been born," *Norway Today,* March 23, 2017, https://nday.info/everyday/scientist-thinks-worlds-first-200-year-old-person-already-born/.

Page 150, *Thus, pandemic-related inventory shortages*: Bobby Marhamat, "Five Consumer Behavior Trends to Watch in 2021," *Forbes,* February 10, 2021, https://www.forbes.com/sites/forbesbusinessdevelopmentcouncil/2021/02/10/five-consumer-behavior-trends-to-watch-in-2021/?sh=5e683bfe3899.

Page 150, *The 2020 Forbes survey we cited earlier: Ibid.*

Page 154, *According to the Associated Press:* Martin Crutsinger, "US productivity in Q4 falls by largest amount in 39 years," Associated Press, February 4, 2021, https://apnews.com/article/health-coronavirus-pandemic-business-1d8cec63d02c037576a90152393322d2.

Page 155, *In studying a wide range of companies:* Jacob Morgan, *The Future of Work* (New Jersey: Wiley, 2014).

Page 158: *With innovation as the most elusive riddle:* Daniel McGinn, "Life's Work: An Interview with Jerry Seinfeld," *Harvard Business Review,* January-February 2017, https://hbr.org/2017/01/lifes-work-jerry-seinfeld.

Page 159, *Meg Whitman, former CEO of eBay:* Jeffrey H. Dyer, Hal Gregersen, and Clayton M. Christensen, "The Innovator's DNA," *Harvard Business Review,* December 2009, https://hbr.org/2009/12/the-innovators-dna.

Page 163, *In this book The Black Swan:* Nassim Nicholas Taleb, *The Black Swan: The Impact of the Highly Improbable,* second edition (New York: Penguin, 2008).

Page 164, *These "pioneers," as author Adam Grant calls them:* Adam Grant, *Originals: How Non-Conformists Move the World* (New York: Penguin, 2016).

Page 166, *A 2019 FlexJobs survey with 7,300 respondents:* Brie Weiler Reynolds, "FlexJobs 2019 Annual Survey: Flexible Work Plays Big Role in Job Choices," FlexJobs, https://www.flexjobs.com/blog/post/survey-flexible-work-job-choices/.

Page 166, *New Republic magazine reports:* "Millennials Don't Want to Climb the Traditional Ladder," *New Republic,* July 30, 2014, https://newrepublic.com/article/118883/millennials-dont-want-climb-traditional-career-ladder.

Page 166, *Sizeable fraction of survey respondents:* Brie Weiler Reynolds, *op. cit.*

Page 167, *There's a lot competing for our attention:* Timothy Egan, "The Eight-Second Attention Span," *New York Times,* January 22, 2106, https://www.ntimes.com/2016/01/22/opinion/the-eight-second-attention-span.html.

Page 168, *The impact on organizations can be painful: How Millennials Want to Work and Live,* Gallup, 2016, https://www.gallup.com/workplace/238073/millennials-work-live.aspx?thank-you-report-form=1.

Page 169, *For example, research reported in the 2018 Digital Transformation Barometer:* "2018 Digital Transformation Barometer," ISACA, https://www.isaca.org/-/media/info/digital-transformation-barometer/index.html.

Page 169, *In 2021, I did a webcast:* Martin Fiore (Producer), 2021, "5 EY Tax Women Alumni Inspire Us to Build a Better Working World—for Women and All."

Page 172, *Soft skills were the most sought-after capabilities:* Cengage, January 16, 2019, "Demand for 'Uniquely Human Skills' Increases Even as Technology and Automation Replace Some Jobs." Press release, https://www.prnewswire.com/news-releases/new-survey-demand-for-uniquely-human-skills-increases-even-as-technology-and-automation-replace-some-jobs-300779214.html.

Page 172, *A similar study by LinkedIn noted: Global Talent Trends 2019,* LinkedIn, https://bsiness.linkedin.com/ctent/dam/me/business/en-us/tlent-stions/rsources/pdfs/global-talent-trends-2019.pdf.

Page 176, *In fact, a Gallup poll released in May 2020:* Jonathan Rothwell and Jessica Harlan, *2019 Gig Economy and Self-Employment Report,* Gallup and Intuit Quick Books, https://quickbooks.intuit.com/content/dam/intuit/quickbooks/Gig-Economy-Self-Employment-Report-2019.pdf.

Page 177, *Tips for effective work-from-home strategies:* Ellen Nastir, "How to Work from Home Effectively: 10 Tips for Working Remotely," Nonprofit Leadership Center blog, March

23, 2020, https://nlctb.org/tips/how-to-work-from-home-effectively/.

Page 182, *An article in Science magazine highlighted:* Matthew Hutson, "Even artificial intelligence can acquire biases against race and gender," *Science,* April 13, 2017, https://www.siencemag.org/news/2017/04/even-artificial-intelligence-can-acquire-biases-against-race-and-gender.

Page 185, *According to the Edelman Trust Barometer:* Edelman Trust Barometer Global Report, Edelman website, https://www.edelman.com/sites/g/files/aatuss191/files/2021-01/2021-edelman-trust-barometer.pdf.

Page 187, *In their 2021 survey, the Edelman Trust Barometer:* Edelman Trust Barometer Global Report, Edelman website, https://www.edelman.com/trust/2021-trust-barometer/insights/employers-must-lead-rebuilding-trust-truth.

Page 190, *In 2019, according to the Oxford Internet Institute:* Azaz Zaman, "A freelancing boom is reshaping Bangladesh and its economy," *The Print,* June 23, 2019, https://theprint.in/world/a-freelancing-boom-is-reshaping-bangladesh-and-its-economy/252971/.

Page 191, *Between 2005 and 2020, the country:* "Indonesia Overview," World Bank website, April 6, 2021, https://www.worldbank.org/en/country/indonesia/overview.

Page 191, *WEF researchers point to programs:* YCAB Foundation website, https://www.ycabfoundation.org.

Page 192, *WEF estimates that up to 1.6 billion learners: The Global Risks Report 2021,* World Economic Forum,

http://www3.wfo-
rum.org/docs/WEF_The_Global_Risks_Report_2021.pdf.

Page 193, *By 2021, India boasted:* "Digital 2021 July Global
Statshot Report," DataReportal website, https://datare-
portal.com/reports/digital-2021-India.

Page 198, *When these issues arise, Linda Goodspeed:* Eve Tah-
mincioglu, "The Board's Role in Setting Up AI's Ethical 'Guardrails,'" *Directors & Boards*, April 5, 2019,
https://www.directorsandboards.com/articles/singleboard's-
role-setting-ai's-ethical-'guardrails'.

Page 211, *In the spring of 2021: Tax Policy and Climate Change:
IMF/OECD Report for the G20 Finance Ministers and Central
Bank Governors,* April 2021, Italy,
https://www.oecd.org/tax/tax-policy/tax-policy-and-climate-
change-imf-oecd-g20-report-april-2021.pdf.

Page 212, *Technology's continued convergence with humanity:*
"How AI Is Helping Us Better Understand the Environment,"
Intel, March 5 2019, https://software.intel.com/con-
tent/www/us/en/develop/articles/how-ai-is-helping-us-better-
understand-the-environment.html.

Page 212, *In a 2018 EY survey, 97 percent* "Nonfinancial disclo-
sures are essential to most institutional investors," EYGM Limited, November 29 2018,
https://www.ey.com/en_jo/news/2018/11/nonfinancial-dsclo-
sures-are-essential-to-most-institutional-investors.

Page 213, *Research shows that ESG efforts:* "The ESG Advantage: Exploring Links To Corporate Financial Performance," S&P Global, April 8 2019,
https://www.spglobal.com/_assets/documents/ratings/the-esg-

advantage-exploring-links-to-corporate-financial-perfor-mance-april-8-2019.pdf.

Page 223, *The concept was first advanced:* Quoted in Mark O'Connell, *To Be a Machine: Adventures Among Cyborgs, Uto-pians, Hackers, and the Futurists Solving the Modest Problem of Death* (London: Granta Books, 2017).

Page 224, *More recently, writers like futurist:* Ray Kurzweil, *The Singularity Is Near: When Humans Transcend Biology* (New York: Penguin, 2006).

Page 225, *According to a study by Weill Cornell Medicine:* Lishomwa Ndhlovu, "In Brief: Researchers Detect Epigenetic Signature of Severe COVID-19," Weill Cornell Medicine news website, February 12, 2021, https://news.weill.cor-nell.edu/news/2021/02/in-brief-researchers-detect-epigenetic-signature-of-severe-covid-19.

Index

About the Author

MARTIN FIORE IS AN EXECUTIVE with Ernst & Young LLP (EY), where he has held a number of key leadership roles. He is a member of the firm's Americas Tax Leadership team and the Americas Inclusiveness Advisory Council, and served as executive sponsor of the technology committee that first introduced intelligent automation at the EY organization.

In addition, Martin is active in organizations related to youth job programs and education and serves on multiple boards of leading universities and charitable organizations. His writing and interviews frequently appear in business and talent media on topics related to tax, the future of business, technology, and talent strategies. He lives on the Upper West Side of New York City with his wife and daughters. Connect with him on Twitter @MartinFioreEY and on LinkedIn.

CPSIA information can be obtained
at www.ICGtesting.com
Printed in the USA
BVHW031006241021
619721BV00012B/102/J